Sport Tourism, Events and Sustainable Development Goals

Discover the transformative potential of sports tourism and events in achieving Sustainable Development Goals (SDGs) with "Sport Tourism, Events and Sustainable Development Goals: An Emerging Foundation." This groundbreaking collection explores the profound impact of these sectors in shaping a more sustainable future.

Readers of this book will gain a deep understanding of how sports tourism and events can serve as powerful catalysts for achieving SDGs. Through a rich array of case studies, innovative strategies, and expert insights, the book provides a roadmap for harnessing the full potential of these sectors to promote social, economic, and environmental sustainability. Readers will benefit from a multidisciplinary approach that integrates theory and practice, offering actionable solutions for scholars, practitioners, and policymakers alike.

This book is intended for scholars, students, professionals, and policymakers interested in the intersection of sports tourism, events, and sustainable development. It serves as an invaluable resource for anyone seeking to leverage the dynamic synergy between these fields to drive positive change and achieve SDGs on a global scale.

Anukrati Sharma is Head & Associate Professor of the Department of Commerce and Management at the University of Kota (a state government university) in Kota, Rajasthan, India.

Miha Lesjak is Associate Professor at the University of Primorska, Faculty of Tourism Studies, Turistica.

Dušan Borovčanin is the CEO of EXPO 2027 Belgrade and the Vice Dean at the Faculty of Tourism and Hospitality Management at Singidunum University in Belgrade and visiting professor at the University of Strasbourg, EM Business School.

Routledge Insights in Tourism Series

Series Editor: Anukrati Sharma
Head & Associate Professor of the Department of Commerce and Management at the University of Kota, India

This series provides a forum for cutting edge insights into the latest developments in tourism research. It offers high quality monographs and edited collections that develop tourism analysis at both theoretical and empirical levels.

Diasporic Mobilities on Vacation
Tourism of European-Moroccans at Home
Lauren B Wagner

Overtourism and Cruise Tourism in Emerging Destinations on the Arabian Peninsula
Manuela Gutberlet

Pseudo-Authenticity and Tourism
Preservation, Miniaturization, and Replication
Jesse Owen Hearns-Branaman and Lihua Chen

Developing Industrial and Mining Heritage Sites
Lavrion Technology and Cultural Park, Greece
Taşkın Deniz Yıldız

Innovation Strategies and Organizational Culture in Tourism
Concepts and Case Studies on Knowledge Sharing
Edited by Marco Valeri

Tourism and Poverty Alleviation in Nature Conservation Areas
A Comparative Study Between Japan and Vietnam
Nguyen Van Hoang

Sport Tourism, Events and Sustainable Development Goals
An Emerging Foundation
Edited by Anukrati Sharma, Miha Lesjak and Dušan Borovčanin

For more information about this series, please visit: www.routledge.com/Routledge-Insights-in-Tourism-Series/book-series/RITS

Sport Tourism, Events and Sustainable Development Goals

An Emerging Foundation

Edited by Anukrati Sharma, Miha Lesjak and Dušan Borovčanin

Routledge
Taylor & Francis Group

LONDON AND NEW YORK

First published 2025
by Routledge
4 Park Square, Milton Park, Abingdon, Oxon OX14 4RN

and by Routledge
605 Third Avenue, New York, NY 10158

Routledge is an imprint of the Taylor & Francis Group, an informa business

ISBN: 978-1-032-47149-5 (hbk)
ISBN: 978-1-032-47150-1 (pbk)
ISBN: 978-1-003-38478-6 (ebk)

DOI: 10.4324/9781003384786

Typeset in Times New Roman
by Apex CoVantage, LLC

Contents

vi *Contents*

Figures

Tables

Contributors

Hristo Andreev is a PhD candidate at Cyprus University of Technology.

Khalid Ballouli is an associate professor and PhD program director, Department of Sport and Entertainment Management, College of Hospitality, Retail, and Sport Management, University of South Carolina, Columbia, SC.

Dušan Borovčanin is an assistant professor at the Faculty of Tourism and Hospitality Management, and the Faculty of Physical Education and Sports Management, the Singidunum University in Belgrade, Serbia.

Miha Bratec is a senior lecturer at the University of Primorska, Slovenia.

Daša Farčnik is an assistant professor at the University of Ljubljana School of Economics and Business, Slovenia.

Hrvoje Grofelnik is a part of the Faculty of Tourism and Hospitality Management, University of Rijeka, Croatia.

Emil Juvan works at the University of Primorska, Faculty of Tourism Studies, Slovenia.

Miroslav Knezevic is a part of the Faculty of Tourism and Hospitality Management, Singidunum University, Belgrade, Serbia.

Kir Kuščer is an associate professor at the University of Ljubljana School of Economics and Business, Slovenia.

Seonjin Lee is a PhD student and research assistant at the Richardson Family SmartState Center for Economic Excellence in Tourism and Economic Development, School of Hospitality and Tourism Management, College of Hospitality, Retail, and Sport Management, University of South Carolina, Columbia, SC.

Miha Lesjak is an associate professor at the University of Primorska, Faculty of Tourism Studies, Turistica, Slovenia.

Lori Pennington-Gray holds an endowed chair and is the director at the Richardson Family SmartState Center for Economic Excellence in Tourism and Economic Development, School of Hospitality and Tourism Management, College of Hospitality, Retail, and Sport Management, University of South Carolina, Columbia, SC.

Marko Perić is a part of the Faculty of Tourism and Hospitality Management, University of Rijeka, Croatia.

Sašo Sever, univ. dipl. ekon. is a Microsoft specialist, lecturer and tutor.

Anukrati Sharma is the head and associate professor of the Department of Commerce and Management at the University of Kota (a state government university) in Kota, Rajasthan, India.

Parag Shukla is an assistant professor at the Department of Commerce and Business Management, The Maharaja Sayajirao University of Baroda.

Terry Stevens is the founder and MD of Stevens & Associates; a visiting professor at The Technological University of the Shannon (Ireland); and a UNWTO advisor on Sports Tourism.

Nicholas Wise works at the Arizona State University, USA.

Preface

As editors of the book "Sport Tourism, Events and SDGs: An Emerging Foundation," it is our distinct pleasure to introduce this comprehensive collection that delves into the dynamic relationship among sports, tourism, events, and the Sustainable Development Goals (SDGs).

Sports, tourism, and events have emerged as formidable forces for positive change in today's global landscape. This book represents a culmination of extensive research and collaboration among leading scholars and practitioners, all of whom share a common goal: to harness the immense potential of these sectors in achieving sustainable development.

In this volume, readers will discover a rich tapestry of insights, case studies, and practical strategies. The benefits of reading this book are manifold. It offers a deep understanding of how these sectors can drive social progress, foster economic development, and promote environmental sustainability. The multidisciplinary approach, coupled with a fusion of theory and real-world applications, should empower scholars, professionals, and policymakers with the knowledge and tools to effect meaningful change.

This book is tailored for a diverse audience – from academics seeking to expand their understanding of this emerging field to practitioners striving to make a positive impact through their work. It is our hope that "Sport Tourism, Events and SDGs: An Emerging Foundation" will serve as a conversation starter for many, inspiring all readers to explore the transformative power of sports, tourism, and events in shaping a more sustainable and inclusive world.

The journey toward achieving the SDGs is a global imperative, and sports, tourism, and events stand as formidable pillars in its realization. We commend the contributions of our esteemed authors and hope that their work will inspire others to join us in this collective mission.

Dr. Anukrati Sharma
Dr. Miha Lesjak
Dr. Dušan Borovčanin

Acknowledgments

The completion of "Sport Tourism, Events and SDGs: An Emerging Foundation" has been a collaborative endeavor, and we wish to express our heartfelt gratitude to the many individuals and organizations that have contributed to the realization of this ambitious project.

First and foremost, we extend our appreciation to the distinguished authors who have shared their expertise and insights. Your dedication to advancing the knowledge in this field has been instrumental in creating a comprehensive collection that will undoubtedly influence and inspire future research and practice.

We are deeply thankful to the reviewers and experts who provided valuable feedback and peer review, ensuring the academic rigor and quality of the content. Your commitment to maintaining the book's excellence is greatly appreciated.

Our gratitude also extends to the editorial and production teams of Routledge, especially to Prachi Priyanka and Leerink Faye, who provided guidance and support throughout the publication process. Their professionalism and attention to detail have been invaluable.

To the scholars, practitioners, and policymakers whose tireless work in the fields of sports, tourism, and sustainable development has inspired this book, we offer our sincere thanks. Your real-world contributions have been a source of inspiration for us and for the readers of this book.

Finally, we would like to acknowledge our families and friends, whose unwavering support and understanding have been the foundation upon which we could embark on this intellectual journey.

We hope that "Sport Tourism, Events and SDGs: An Emerging Foundation" will serve as a meaningful resource for scholars, students, professionals, and policymakers seeking to harness the transformative power of sports, tourism, and events in the pursuit of a more sustainable and inclusive world. It is through the collective efforts of all those mentioned earlier that we have been able to bring this vision to life.

With deep gratitude,

Dr. Anukrati Sharma
Dr. Miha Lesjak
Dr. Dušan Borovčanin

1 Sports tourism and Sustainable Development Goals

Foundations and pathways

Parag Shukla, Pankaj Kumar Tripathy, and Jahanvi Bansal

Introduction

Sports tourism has emerged as a multi-dollar industry in recent years, with a current valuation of $323 billion in 2020 and a potential to reach $1,803 billion by 2030 (Kadam & Deshmukh, 2021). The industry is expected to witness an annual growth rate of 16.1% from 2021 to 2030. India is no exception to this, as its sports tourism industry has robustly carved a niche position for itself in recent decades.

India is an incredible travel destination that offers a plethora of experiences including spiritual, mystic, and escapades to tourists. India's historical, topographical, and geographical diversities have paved several roads for the use of sports as a tourist venture. Inevitably, sports tourism has notched a prominent and promising place in India's tourism industry (Mohite & Bhosale, 2017). According to a report by Future Market Insights (FMI), India's sports tourism is an exploding market and is projected to surge by $376 billion by 2032. The industry is even immune from the post-pandemic effects, as stated by an expert from FMI (Chanda, 2023). The underlying reasons may be attributed to the notion that people are becoming more health-conscious post-pandemic and, so, are expected to actively participate in sporting activities. To boot, owing to people's never-ending passion for sports, sports tourism has regained momentum following the lifting of travel restrictions across the globe (Kadam & Deshmukh, 2020).

Throughout history, humans have had an incessant urge and inclination to travelling for sports (Kurtzman & Zauhar, 2003). Perhaps, there is no doubt that tourism has been, and maintains to be, a focal source of development (Zauhar, 2004). Nations view sports tourism as a growth strategy to revitalize social, economic, technological, and environmental development (Emery, 2002; Larissa, 2010; Yusuf, 2017). However, any action principally focusing on the current situation is now considered dubious and obsolete. Recent investigations have bid for the obligatory addition of 'sustainable' as a prefix in the development terminology (Panagiotopoulos et al., 2022). Despite several empirical pieces on sports tourism, a strong theoretical foundation is missing from its line of research, which could explain its antecedents and consequences in a single framework and establish a connection with 17 Sustainable Development Goals (SDGs) articulated by the United Nations in 2015.

DOI: 10.4324/9781003384786-1

There is a pressing need to build a conceptual framework that explores the motives that drive individual motivation toward sports tourism and its bearing on 17 SDGs via society, which, as a living system, operates on the five hierarchical levels of motivation/needs given in Maslow's pyramid (Panagiotopoulos et al., 2022). A burgeoning body of literature has analyzed travelers' needs and motivations by linking them with Maslow's need hierarchy theory (Kurtzman & Zauhar, 2003; Yousaf et al., 2018), including the travel career ladder (TLC) proposed by Pearce (1988) and travel career patterns (TCP) suggested by Pearce and Lee (2005). This need for applying theories from parent disciplines, including sociology or social psychology, to get the progression of the knowledge base and development of a comprehensive understanding of sports tourism was also supported by Gibson (2003). Nevertheless, the current study fills the gap in the literature by associating Maslow's theory with the contemporary and sustainable needs of society, all other things considered. This way, this study is one of a kind, providing a multidisciplinary integrative model depicting the relationship between sports tourism and 17 SDGs, considering individual and societal motives, respectively.

This study has been descriptive, and the data for this study were obtained from secondary sources, including papers, manuscripts, articles, news, and books published on the selected theme. The authors have attempted to do an in-depth and rigorous reading of all the carefully chosen references, which will constitute the foundation of this review study.

Literature review

Sports tourism

The term sports tourism has been coined to flash and flaunt tourism exclusively based on the theme of sports. Over the period, it has gained noteworthy attention from the field of researchers and academicians. One school of thought specifies sports tourism as traveling beyond conventional reasons to either engage in or view sports-related activities. While the former is known as 'active' sports tourism, the latter is termed as 'passive' sports tourism (Haldar, 2017). Several other researchers have propagated the categorization of sports tourism. The most widely acclaimed and overarching is the one given by Gammon and Robinson (1977). They, on the basis of travelers' motives, asserted an interesting interdisciplinary exchange of sports and tourism broadly classified into 'sports tourism' and 'tourism sports.' Travelers giving priority to sports activities over visits come under the former category, while the ones more interested in visiting sports sites and events come under the latter. They further opined that sports tourism could also be categorized as 'hard' and 'soft' tourism sports. Parallel to the above categorizations, 'hard' sports tourism includes travelers traveling for either active or passive participation in competitive sports events. In contrast, 'soft' sports tourism is meant for active participation in recreational sports, including running, canoeing, skiing, and hiking.

In a slightly different view, there is an ongoing trend of 'celebrity and nostalgia' sports tourism that revolves around visiting famous sports museums, halls, and venues and sometimes meeting iconic sports personalities (Haldar, 2017). Another emerging view is combining business trips with recreational sports activities to keep employees entertained and engrossed. Not least of all, professional sportspersons have nurtured sports tourism by widely traveling for their career (Mohite & Bhosale, 2017).

Sports tourism and SDGs

Irrefutably, there are notable positives of sports tourism including the promotion of local culture, aiding economic and social development, strengthening national heritage, and deepening community spirit (Chavan, 2020). A probe into the area shows evidence that the lack of planning and ignorance about sustainability have made tourist destinations sheer victims of its success (Jiménez-García et al., 2020).

Owing to this, the concepts of sustainable development and growth in the tourism sector took a formal turn in 2015 during the general debate on sustainability led by the United Nations. This resulted in the 2030 Agenda grounded on sustainable development, which includes 17 SDGs (Figure 1.1), each specifically targeting the sustainable needs of the planet and society (Jiménez-García et al., 2020). This confirms and validates a close link between sports tourism and sustainability explored in this study.

Figure 1.1 depicts a conceptual framework that underlines the purpose of this study in two folds. First, this study aims to identify the motives behind an individual's sports tourism intention, thereby allowing a better understanding of the antecedents of the current research theme. Second, this study offers a strong theoretical foundation that could explain the consequences of sports tourism while establishing a connection between the application of Maslow's need hierarchy theory in society and 17 SDGs at different hierarchical levels of the said theory.

Individual motives and sports tourism intentions

Human actions can be understood by determining the underlying motives, drives, and concerns (Zauhar, 2004). This calls for the understanding of the driving motives behind individual participation in sports tourism. Hemmatinezhad et al. (2010) argued that discovering individual drivers for joining sports-based tourism is the stepping stone for fostering a theoretical understanding of how to make the most out of this industry or, in our case, to achieve SDGs. Aicher and Brenner (2015) used self-determination theory (SDT), developed by Deci and Ryan (1985, 2000), to outline the overlapping linkage between the controlled and autonomous motivators influencing individual sports participation decisions. According to Crompton and McKay (1997), 'tourism motivation is conceptualized as a dynamic process of internal psychological factors (needs and wants) that generate a state of tension or disequilibrium within individuals.'

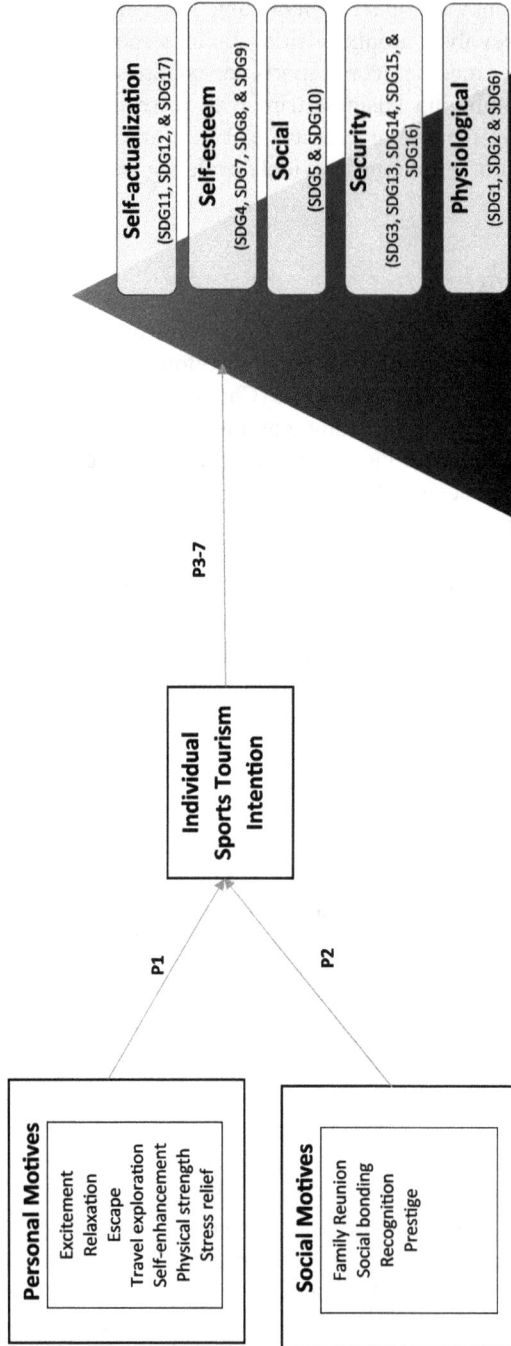

Figure 1.1 Conceptual framework of the impact of sports tourism on SDGs based on Maslow's need hierarchy theory.

Sports tourism psychology states that people are usually motivated by several drivers and that another may reinforce one motivator. A mounting body of literature has put forward several other tourism motivators, including knowledge, social and prestige motivation (Baloglu & McCleary, 1999), excitement, relaxation, reunion with family and friends, escape from daily life and stress, fun while ensuring safety, and enjoying the natural landscapes (Yoon & Uysal, 2005). In a recent cross-national study on participation motivation in active sports tourism, Mishra, Malhotra et al. (2022) categorized motives as 'personal,' including exploration, self-enhancement, and stress release, and 'social,' comprising social bonding and recognition. Specifically, the current study has utilized previous research to explain individuals' motives that gear them to actively engage in sports tourism and events. However, it is important to note that most of the research outlined earlier has concentrated on active sports tourists. Therefore, this chapter will focus on all the categories of sports tourists as explained in the above section to get a comprehensive view of the impact of individual sports tourism motivation. Based on the above literature review, the present study is categorized into personal and social motives. Hence, the following hypotheses are proposed.

P1: Personal motives will have a positive impact on individual sports tourism intentions.

P2: Social motives will have a positive impact on individual sports tourism intentions.

Sports tourism and application of Maslow's theory to SDGs

Sports tourism, physiological needs of society, and SDGs

Maslow's first level of needs, that is, 'physiological needs,' is associated with an individual's basic needs for livelihood, such as food, shelter, and water. Applying the same impression to society, physiological needs refer to the basic needs of society. These needs can be capitalized on by sports tourism and will directly affect the three SDGs, namely, SDG 1 'No poverty,' SDG 2 'Zero hunger,' and SDG 6 'Clean water and sanitation.'

Even though battling hard with poverty, India is still home to the highest number of poor people globally. Unsurprisingly, India is ranked 107th among the 121 countries included in the Global Hunger Index (GHI) 2022 (Roy, 2022). Sports tourism, like other tourism sectors, provides several direct and allied sources of income even to less favored groups. This helps the nation in pushing back against poverty and hunger from society. The substantial contribution of sports tourism to the nation's economy is a clear indicator of the reduction of poverty and hunger in regional areas backed by sports tourism (Panagiotopoulos et al., 2022). Sourcing food from sustainable and responsible local producers and making responsible food consumption decisions are new trends observed by several sports event organizers and other stakeholders (Sport for Sustainable Development, 2020). Sports tourism has played a prominent role in the fight for water cleaning and sanitation, not only to provide customer experience to tourists but also because of the need of the hour coupled with the awareness of the theme. Another striking reason behind the development of the localities for clean water and sanitation is the hosting of international sports events such as the Commonwealth Games, Olympic Games, or World Championships. Hence, the following propositions are developed:

P3: Sports tourism motivation will have a positive impact on the physiological needs of society in the context of SDGs.
P3a: Sports tourism motivation will have a positive impact on SDG 1.
P3b: Sports tourism motivation will have a positive impact on SDG 2.
P3c: Sports tourism motivation will have a positive impact on SDG 6.

Sports tourism, security needs of society, and SDGs

The second need purported by Maslow is 'security,' derived from individuals' apprehension for a safe and secure life. This study explores the impact of sports tourism on the security needs of society by relating to five SDGs, namely, SDG 3 'Good health and well-being,' SDG 13 'Climate action,' SDG 14 'Life below water,' SDG 15 'Life on Earth,' and SDG 16 'Peace, justice, and strong institutions.'

Medical research has shown that sports considerably lessen the probability of heart disease, stroke, diabetes, and, in some cases, cancer. This ensures healthy lives and promotes the well-being of the members of society. Green sport is an emerging field aiming to organize climate-friendly sports activities and events (Panagiotopoulos et al., 2022). Darmawan et al. (2022) stressed that water-based sports tourism is responsible for minimizing its negatives and conserving marine life while nurturing itself through community participation.

Sports and the environment are correlated in the sense that the existence of one has a pertinent impression on the other. With sports tourism relying on a healthy natural environment, it is equally imperative to see that the domain is not harmed. Sensibly all stakeholders of sports tourism should take steps to mitigate the adverse biodiversity effects, including damage to natural areas, trampling of vegetation, and disturbance of sensitive species, to name a few (International Union for Conservation of Nature, 2022). Furthermore, sports tourism eases interactions between people of different nations and, so, fosters a sense of mutual acceptance and understanding (Sport for Sustainable Development, 2020). This lays the foundation of robust institutions and peaceful societies. So, the following propositions can be advanced:

P4: Sports tourism motivation will have a positive impact on the security needs of society in the context of SDGs.
P4a: Sports tourism motivation will have a positive impact on SDG 3.
P4b: Sports tourism motivation will have a positive impact on SDG 13.
P4c: Sports tourism motivation will have a positive impact on SDG 14.
P4d: Sports tourism motivation will have a positive impact on SDG 15.
P4e: Sports tourism motivation will have a positive impact on SDG 16.

Sports tourism, social needs of society, and SDGs

The third need of the hierarchy emphasizes developing a sense of social belonging and healthy relationships among the members. This framework is studied through two SDGs, namely, SDG 5 'Gender equality' and SDG 10 'Reduced inequalities.'

Sports predominantly witness the male–female divide. In many societies, sportswomen are perceived as 'masculine,' while men uninterested in sports are judged as 'unmanly.' Sports tourism is helping society in breaking the stereotypes attached to gender in several ways (Apollo et al., 2023). With more women opting for sports vacations, women-only travel groups are up and coming, thus generating employment avenues for local women. Female athletes active in sports tourism also act as role models for local women. Sports tourism, especially events, can pass on the message of equality and inclusivity to the masses while addressing the issue of discriminatory practices (Sport for Sustainable Development, 2020). Parasport activities give platforms to different abled athletes and, at the same time, can be enjoyed by sports tourists. The following statements can be proposed:

P5: Sports tourism will have a positive impact on the social needs of society in the context of SDGs.
P5a: Sports tourism will have a positive impact on SDG 5.
P5b: Sports tourism will have a positive impact on SDG 10.

Sports tourism, self-esteem needs of society, and SDGs

The fourth motivation of the cited author's theory comprises self-esteem and personal development. Nevertheless, in the context of the present study, it embraces the development of society as a whole using SDG 4 'Quality education,' SDG 7 'Affordable and clean energy,' SDG 8 'Decent work and economic growth,' and SDG 9 'Industry, innovation, and infrastructure.'

Meaningful training in sports tourism using modern techniques coupled with experimental learning programs facilitates considerable upgradations in the level of the studied quality (Demyanova et al., 2021). Sports-based programs offer employment opportunities and transversal life skills to be used beyond school. Contrary to this, local people deprived of elementary education can work as instructors and guides and are often likely to get paid more than average for their skills. Residents can make decent and productive work lives through allied businesses such as local hospitality businesses, equipment rentals, and transport services. SDG 7 aims to offer reasonably priced, sustainable, and modern energy for all. Building new energy-saving stadiums and other sports facilities while re-structuring and engineering existing ones is imperative for achieving the said goal (Sport for Sustainable Development, 2020). As an indispensable part of the national industry, the sports industry has the potential to build a state-of-the-art infrastructure while fostering sustainable industrialization and innovation. For instance, using augmented reality- (AR-) and virtual reality- (VR-) based wearable devices can heighten the customer's experience multiple times and make sports tourism more exciting. Thus, the following propositions are advanced:

P6: Sports tourism motivation will have a positive impact on the self-esteem needs of society in the context of SDGs.
P6a: Sports tourism motivation will have a positive impact on SDG 4.
P6b: Sports tourism motivation will have a positive impact on SDG 7.
P6c: Sports tourism motivation will have a positive impact on SDG 8.
P6d: Sports tourism motivation will have a positive impact on SDG 9.

Sports tourism, self-actualization needs of society, and SDGs

The ultimate need in the hierarchy is 'self-actualization.' Self-actualization about society relates to developing a self-propelling mechanism leading to sustainable growth over the future centuries. The same is proposed to be achieved through SDG 11 'Sustainable cities and communities,' SDG 12 'Responsible consumption and production,' and SDG 17 'Partnerships for the goals.'

Rapid urbanization and our unpreparedness to appropriately deal with it have led to several downsides, such as intensified air pollution, inadequate access to public transport, and diminished safety and resilience to climatic changes and natural disasters. Sports tourism presents a powerful platform for raising awareness of the threats posed to the existence of urban settlements. This depicts a significant step toward creating sustainable cities and communities. Equally applicable are the sustainable practices that aim to reduce, reuse, and recycle. Undeniably innovative and responsible consumption and production habits will lessen future economic, social, and environmental costing (Sport for Sustainable Development, 2020). In closing, establishing partnership goals at local, national, and international levels is essential for the above-outlined self-propelling mechanism. This calls for a joint venture among varied stakeholders of sports tourism, including policymakers, sports clubs, and associations; event organizing companies; the corporates; national and international sports federations; media; sports celebrities; and academicians and researchers. On the basis of the above, the following propositions are made:

P7: Sports tourism motivation will have a positive impact on the self-actualization needs of society in the context of SDGs.
P7a: Sports tourism motivation will have a positive impact on SDG 11.
P7b: Sports tourism motivation will have a positive impact on SDG 12.
P7c: Sports tourism motivation will have a positive impact on SDG 17.

Managerial implications and conclusions

For many decades sports has been exhibiting a pertinent role in society. Here, the concept of sustainable development aptly fits into it. Sports' universal likeability can act as a catalyst in energizing the agenda of SDGs. Discovering this enormous potential of sports and subsequently sports tourism, this study offers several policy implications. The first is a powerful partnership among planning agencies, tourism stakeholders, corporates, and research universities to build information-sharing programs aimed at the realization of the SDGs at different need-based levels of society. Second, conducting scientific cause–effect studies to assess ecological impacts, launching educational awareness programs for local people and communities, and communicating the message of world peace to provide a safe environment at the grassroots level can also be pivotal in this direction. Third, the government should introduce subsidies for sports organizers who hire local people, encourage local entrepreneurs, and utilize local products to unlock the potential of sports tourism at a local level. Fourth, as a change agent for society, sports tourism can encourage social inclusiveness with local practices, customs, and needs, taking all into consideration. For instance, uplifting accessible sports tourism for people with disabilities and promoting balanced participation of genders in sports events can foster social development despite significant barriers for the specified categories. Fourth, sports tourism organizers should make sincere efforts to take eco-friendly measures, including the reuse and recycling of sports utilities and restricting the use of hazardous materials (e.g., pesticides) and vehicles emitting harmful gases to preserve people and the planet. Finally, a research agenda should be propagated to analyze and understand sports tourists' motives and behaviors to ensure low-impact tourist visits.

References

Aicher, T. J., & Brenner, J. (2015). Individuals' motivation to participate in sport tourism: A self-determination theory perspective. *International Journal of Sport Management, Recreation and Tourism*, 18(1), 56–81.

Apollo, M., Mostowska, J., Legut, A., Maciuk, K., & Timothy, D. J. (2023). Gender differences in competitive adventure sports tourism. *Journal of Outdoor Recreation and Tourism*, 42, 100604.

Baloglu, S., & McCleary, K. W. (1999). A model of destination image formation. *Annals of Tourism Research*, 26(4), 868–897.

Chanda, K. (2023). *Fun and games: How sports tourism is picking up pace in India*. Retrieved on 10th March, 2023 from www.forbesindia.com/article/take-one-big-story-of-the-day/fun-and-games-how-sports-tourism-is-picking-up-pace-in-india/81117/1

Chavan, U. (2020). Benefits of sports tourism in current scenario. *Aayushi International Interdisciplinary Research Journal (AIIRJ)*, 3(1), 137–139.

Crompton, J. L., & McKay, S. L. (1997). Motives of visitors attending festival events. *Annals of Tourism Research*, 24(2), 425–439.

Darmawan, R., Abidin, J., & Widyaningsih, H. (2022). *Development of water sport tourism based on sustainable tourism in Pramuka Island, Thousand Islands DKI Jakarta*. 5th International Conference on Sport Science and Health (ICSSH 2021). Dordrecht, The Netherlands: Atlantis Press, 153–157.

Deci, E. L., & Ryan, R. M. (1985). *Intrinsic motivation and self-determination in human behavior*. New York: Plenum.

Deci, E. L., & Ryan, R. M. (2000). The "what" and "why" of goal pursuits: Human needs and self-determination of behavior. *Psychological Inquiry*, 11(4), 227–268.

Demyanova, L., Usova, I., Tashchiyan, A., Ryzhkin, N., & Demyanov, S. (2021). *Role of sports tourism in the school children health improvement*. Rostov-on-Don, Russia: E3S Web of Conferences 273, 09037.

Emery, P. R. (2002). Bidding to host a major sports event: The local organization committee perspective, *International Journal of Public Sector Management*, 5(4), 316–335.

Gammon, S., & Robinson, T. (1997). Sport and tourism: A conceptual framework. *Journal of Sport Tourism*, 4(3), 11–18.

Gibson, H. J. (2003). Sport tourism: An introduction to the special issue. *Journal of Sport Management*, 17(3), 205–213.

Haldar, D. (2017). "Sports tourism": Impact and aspect in India. *International Journal of Physical Education, Sports and Health*, 4(4), 292–297.

Hemmatinezhad, M. A., Nia, F. R., & Kalar, A. M. (2010). The study of effective factors on the motivation of tourists participating in sport events. *Ovidius University Annals, Series Physical Education & Sport/Science, Movement & Health*, 10(2), 356.

International Union for Conservation of Nature. (2022). *IUCN and Sails of Change launch sports for nature partnership to help sports become nature-positive*. Retrieved on 5th March, 2023 from www.iucn.org/news/202210/iucn-

Jiménez-García, M., Ruiz-Chico, J., Peña-Sánchez, A. R., & López-Sánchez, J. A. (2020). A bibliometric analysis of sports tourism and sustainability (2002–2019). *Sustainability*, 12(7), 2840.

Kadam, A., & Deshmukh, R. (2021). Sports tourism market by product (football/soccer, cricket, motorsport, tennis, and others), type (domestic and international), and category (active and passive): Global opportunity analysis and industry forecast 2021–2030. *Allied Market Research*, 245–275.

Kurtzman, J., & Zauhar, J. (2003). A wave in time – the sports tourism phenomena. *Journal of Sport Tourism*, 8(1), 35–47.

Larissa, D. (2010). Sport and economic regeneration: A winning combination. *Journal of Sport in Society*, 13(10), 1438–1457.

Mishra, S., Malhotra, G., Johann, M., & Tiwari, S. R. (2022). Motivations for participation in active sports tourism: A cross-national study. *International Journal of Event and Festival Management*, 13(1), 70–91.

Mohite, S. B., & Bhosale, S. V. (2017). Sports tourism in India: An overview. *International Journal of Researches in Social Sciences and Information Studies*, 1, 50–51.

Panagiotopoulos, P., Mitoula, R., Georgitsoyanni, E., & Theodoropoulou, E. (2022). The contribution of sports tourism to sustainable development based on sustainable development indicators – the case of Greece. *Journal of Interdisciplinary and Multidisciplinary Research*, 5(7), 1666–1678.

Pearce, P. L. (1988). The Ulysses factor: Evaluating tourists in visitor's settings. *Annals of Tourism Research*, 15(1), 1–28.

Pearce, P. L., & Lee, Uk-Il. (2005). Developing the travel career approach to tourist motivation. *Journal of Travel Research*, 43(3), 226–237.

Roy, E. (2022). *India ranked 107 on hunger index; govt says bid to taint country's image*. *The Indian Express*. Retrieved on 5th March, 2023 from https://indianexpress.com/article/india/global-hunger-index-india-2022-8209687/

Sport for Sustainable Development. (2020). Score all 17. *Erasmus and European Union*. Retrieved on 5th January, 2023 from http://sport4sd.com/

Yoon, Y., & Uysal, M. (2005). An examination of the effects of motivation and satisfaction on destination loyalty: A structural model. *Tourism Management*, 26(1), 45–56.

Yousaf, A., Amin, I., & C Santos, J. A. (2018). Tourist's motivations to travel: A theoretical perspective on the existing literature. *Tourism and Hospitality Management*, 24(1), 197–211.

Yusuf, T. G. (2017). Impact of sport tourism on the host community: A case study of Nigerian Universities' games. *Journal of Tourism Research*, 18(1), 105–121.

Zauhar, J. (2004). Historical perspectives of sports tourism. *Journal of Sport & Tourism*, 9(1), 5–101.

2 What lies in the state of art of sports and SDGs? Analysing past, present, and future for sustainability

Manpreet Arora and Monika Chandel

Introduction

According to the United Nations, sports is a vital promoter and facilitator for achieving Sustainable Development Goals (SDGs). Sports has great power for transforming the world. The United Nations' 2030 Agenda for achieving development recognizes sports as an important driver and enabler for sustainable development. Sports has a great ability in contributing towards the development of peace and promotion of tolerance all around the world. There are multiple sustainable development goals that can be achieved through sports. An important one is addressing the partnerships all around the world. Sports not only provides various social benefits but it also instils and promotes healthy lifestyle choices among masses. It can help combat various non-communicable diseases and promotes the overall well-being of the society. Sports has been considered to be a great driver in promoting gender equality, where men and women are treated equally. It helps raise and build self-esteem and also promotes various levels of skills so that men and women become equal participants in their various communities. The true spirit of sports lies in social inclusion, which has a great ability to promote awareness within every kind of society. Those who are vulnerable or people with various disabilities are also included in sports, and the objective of providing equal opportunity to all is achieved. Through sports, a lot of public and private partnerships can be fostered, which drives the achievement of SDGs through the contribution of various stakeholders. Governments, private sector organizations, public sector organizations, and local communities all come together when any sports event is organized, thereby leading to greater achievement of SDGs.

Sports also promotes various entrepreneurial activities, and it has the potential to create jobs. It is a great avenue of sustainable income and also promotes decent work and life for all. Through sports and various sports events that are organized at national and international levels, all the communities come together, and this promotes various sustainable activities on the part of stakeholders. For example, if we talk about any World Cup or Asian Games or Olympics, the whole world comes together to celebrate the spirit and true sportsmanship of various players from various countries. It also promotes tourism to a great extent and is a great enabler of fostering tourism activities across the world. The service sector also benefits a lot from the sports sector. When there are sports events, medical tourism, the service

DOI: 10.4324/9781003384786-2

sector, tourism, and hospitality all get a boost. It has been regarded as a very pow-erful tool for promoting human values, ethics, teamwork, discipline, diversity, and empathy across the various communities. It is regarded to have the capabilities of inducing social cohesion, and when these social values are instilled into young peo-ple, they become better citizens, and they promote peace. All the problems related to religion, tribes, and ethnic differences are kept aside because the love of the sport and the sportsmanship are kept above everything.

Sports is used as a vehicle for overall development and peace building across the nations. It has a great ability to contribute towards various national objectives and SDGs. Sports is a great facility that can be used to contribute towards improving health and education, creating employment opportunities, bringing about economic development, and realizing gender equality. Sports also helps stimulate innova-tions at various levels by creating new ideas, techniques, and approaches for value creation of the humanity as a whole. Most of the sports have seemed to be having global impacts; therefore, various innovations that come under sports are helping in lifting the barriers of participation and bringing the nations together at a com-mon place. According to the United Nations, it helps achieve various sustainable goals, and it is an important contributor towards the realization of development and peace goals. It is an effective mechanism that brings equal opportunities for women, young people, and religious communities; promotes health, education, and social inclusion objectives; and promotes tolerance and respect for all.

Methodology

The present chapter is qualitative and quantitative and is majorly literature based. Through the Scopus database, 96 initial documents were retrieved using the search string *"sports" AND "sustainable development goals" OR "SDG"* on 22 March 2023 at 10.40 am. Only the article and review documents written in the English lan-guage were taken into consideration and published in journals only. Both the final article and the article in press were taken for analysis. This refined research produced 78 documents, which are further utilized for satisfaction of research questions.

Research questions

RQ1: How far are sports and SDGs interrelated with each other?
RQ2: What are the areas of research where sports and SDGs are aligned?
RQ3: What is the future scope of the field of sports for achieving SDGs?

In order to find out the research being done on sports and the achievement of SDGs, the authors' contribution was determined by conducting a thematic analysis of the database. Highly cited documents were also figured out to understand the research scope of the area. Keyword analysis was also performed to identify the niche areas of the themes discussed, explored, and contributed by the academic community.

To answer the research questions, two software packages VOSviewer and R-Studio Bibliometrix were used to analyse the search results obtained from the Scopus database.

Sports and SDGs – the niche areas of research

In order to understand the future of the achievement of SDGs through sports, it is necessary to ponder upon research already done in the said area. The academic community has been discussing, researching, and highlighting the importance of sports in achieving SDGs through their research articles. The current dataset highlights various important niche areas in relation to sports and SDGs. For this purpose, VOSviewer software was used to assess the co-occurrence of the authors' keywords. Figure 2.1 generated through VOSviewer software highlights the research focus with the help of clusters. The co-occurrence analysis highlights the research focus with the help of eight clusters. These clusters indicate various niche areas of research done on "sports and SDGs". It also indicates the developed or less developed areas of research, which can help us in forming the future agenda of research. Cluster 1 clearly highlights the contribution of sports towards Agenda 2030. It highlights the importance of physical education (PE) and physical activity (PA) at school and other levels for the achievement of SDGs. One aspect of this research cluster highlights the need of focusing on governance issues also. Cluster 2 also highlights the importance of sports for development and sustainable growth. It talks about the need for innovation in this area. Furthermore, the steps taken by the emerging nations and the less developed nations in this direction are discussed by academicians in this research cluster.

Sports can contribute a lot to achieving the SDGs related to health, education, gender equality, and overall development. The United Nations is playing a pivotal role in this direction. Research Cluster 3 highlights the research contribution in these areas. There are also certain challenges that need to be addressed in relation to sports and SDGs at the policy levels. Cluster 4 highlights the importance of policy intervention in this direction for achieving SDGs. The main areas of highlight where policy framing is necessary are health, higher education, and sports diplomacy. In general parlance, it is assumed that sports contribute towards "peace and harmony".

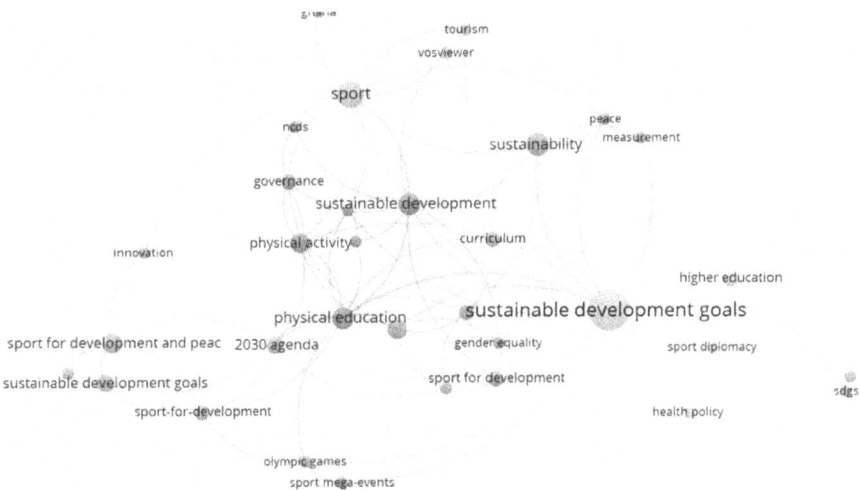

Figure 2.1 Co-occurrence of the authors' keywords.

Cluster 5 indicates this aspect of sports where research contributions talk about "peace and harmony". Cluster 6 indicates the importance of sports in promoting tourism. Various emerging destinations are encouraging their tourism sector with the help of sports events and sports festivals. There is great scope of research in the area of sports tourism, and this fact is supported by this cluster in this dataset. The circles of the cluster are very small and the key thrust areas are very less, indicating a lot of research scope in the area of "Sports tourism and achievement of SDGs".

The next cluster, which is cluster 7, highlights the concept of social sustainability and SDGs. The concept of "social sustainability and inclusion" emphasizes the importance of putting people in priority while talking about the development process. It highlights the need of understanding and manages the negative effects of development on the poor and deprived sections of society. We must understand the importance of creating "resilient societies". While we travel the road of growth and development, we must not forget the poor and vulnerable and should care about creating a cohesive and healthy environment for the deprived sections of society. Cluster 8 highlights the global impact of organizing sports mega events like the Olympics, Commonwealth Games, and World Cups in different sports categories. There are certain world expositions that have the capability of reaching global audiences. These events have shown various deeper social impacts, not only on the host country but also on other parts of the world. These mega events can achieve multiple SDGs like promoting peace and harmony, encouraging global partnerships, inducing gender equality, and fostering overall well-being and overall economic development of a country. There is a great scope for researchers to understand the social, economic, and sustainable impacts of these sports mega events on various sectors of the economy like entrepreneurship, service sector, food and hospitality, transportation, tourism, infrastructure, media and journalism, and education.

Most frequent words

Figure 2.1 depicts the most trending topics of the dataset extracted from the Scopus database. The most frequent keywords, that is, "sustainable development goals", "sports", "sustainability", "physical education" and "sustainable development", highlight the key research areas that are the most trending topics of the research in the current scenario. This indicates that research in this area is on the boom and will continue to be the focus of research in the area of sports and SDGs in the future.

If we look at the words that are most frequently used in the dataset on "sports and SDG research", we get our overview about the trends in research and future research agenda also. In addition to the main keywords, that is, "sustainable development goals" and "sports and sustainability", one important aspect that comes out in the dataset is that academicians are discussing "physical education". It highlights that in the coming years, in addition to sustainability and SDGs, the thrust of academicians is going to be on "physical education" and the related dimensions of "sustainable development". Sports has come out to be an important driver of "sustainability and sustainable development", in which "physical education" and development in the areas of PE are going to be very important in the years to come. It also highlights certain policy implications that if governments want to explore sports as an area of sustainability and sustainable development, then they have to put a thrust on PE and bring out norms and policies related to fostering PE in each country.

Content analysis of the dataset has been done further to understand the relation between sports and SDGs and the focus of sports. The word cloud in Figure 2.3 highlights the most frequent words used by researchers in their articles. In addition to the main keyword, that is, sports, the major highlights or the focus of research is on "sustainable development goals", "sustainable development",

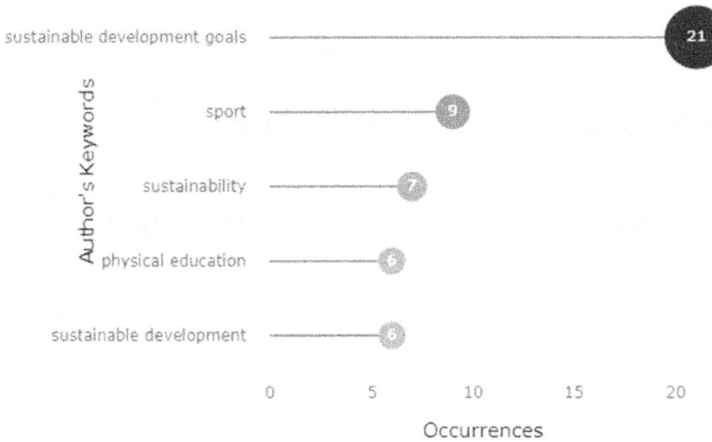

Figure 2.2 Words most frequently used in sports and SDG research.

Figure 2.3 World cloud of the most frequently used keywords.

"physical education", "physical activity", "sports for development", "gender equality", "innovation", "2030 Agenda", "sports mega events", "sports in curriculum", "peace", "governance", "policies", and "higher education".

Highly contributing articles

Research in the area of sports and SDGs is at a very nascent stage, and very few contributions can be seen in the literature, especially in the Scopus database. The authors have attempted to analyse the top-cited 15 papers on sports and SDGs in order to create a wider perspective about the niche areas of research in these areas. Table 2.1 shows the 15 highly cited articles. In addition to these papers, the themes and keywords of the articles are also mentioned. Furthermore, these papers have contributed towards the SDGs mentioned in Table 2.1. It could be seen from the table that the research, which is top cited, majorly talks about Goals 8, 12, 3, 17, 4, 15, 16, 9, and 13. The major thrust of all the highly cited articles in one way or another lies on sustainable development and peace. And they also talk about the related dimensions such as quality, education, clean energy, partnerships for the goals, good health and well-being, and gender equality.

Jiménez-García et al. (2020) explained sports tourism in relation to sustainability and analysed this research lens from different perspectives. The results showed that participation at the community level is essential to achieve the 2030 Agenda of SDGs through sports tourism. The evolution of both fields has led to their interconnectedness, which deals with sustainability as a cross-cutting issue. They further identified major upcoming research issues highlighting the importance of education and destination planning with a special emphasis on sports tourism and SDGs. Lindsey and Darby (2019) highlighted sports as "an important enabler" to attain Agenda 2030. Their analysis focused on three specific orientations. The first one is the sports for development and peace "moment" towards SDG 4, which mainly focuses on education and related issues. The second one is "potential synergies between sports participation policies" and SDG 3, which addresses "health and well-being targets reducing non-communicable disease". The last one is practices within professional football in relation to several migration-related targets of SDGs. They highlighted that policy intervention is essential for achieving SDGs.

Baena-Morales et al. (2021) in their research focused on specific SDGs with special reference to PE and then related those specific SDGs with different models based on PE practices. This was done with special regard to teachers in the field of PE so that they can spread awareness among students for a more sustainable world. Agenda 2030 in relation to education was emphasized in this paper. Their research indicated that focusing on improving the PE curriculum for sports can facilitate some of the SDGs such as

SDG 3 on good health and well-being, SDG 4 on quality education, SDG 5 for bringing gender equality, SDG 8 which emphasises economic growth and decent work, SDG 10 on reducing inequalities, SDG 11 sustainable communities, SDG 12 for responsible production and consumption, SDG 13 addresses climate action, SDG 16 on peace and justice and lastly SDG 17 on partnership for all goals.

Table 2.1 Top 15 highly cited articles

Sr. no.	Title	Author	Journal	TC	DOI
1	A bibliometric analysis of sports tourism and sustainability (2002–2019)	Jiménez-García et al. (2020)	*Sustainability (Switzerland)*	53	10.3390/su12072840
	Themes and keywords of the article	Research trends, SciMAT, sport, sustainable, tourism, VOSviewer			
	SDGs mapped to this document	Goal 4 = Quality education, Goal 8 = Decent work and economic growth, Goal 12 = Responsible consumption and production			
2	Sport and the Sustainable Development Goals: Where is the policy coherence?	Lindsey In (2019)	*International Review for the Sociology of Sport*	47	10.1177/1012690217752651
	Themes and keywords of the article	2030 Agenda, education, health, migration, United Nations			
	SDGs mapped to this document	Goal 3 = Good health and well-being, Goal 4 = Quality education, Goal 12 = Responsible consumption and production, Goal 17 = Partnerships for the goals			
3	Sustainable development goals and physical education. A proposal for practice-based models	BAENA-MORALES S et al. (2021)	*International Journal of Environmental Research and Public Health*	31	10.3390/ijerph18042129
	Themes and keywords of the article	Pedagogical research, physical activity, physical education, sustainability development			
	SDGs mapped to this document	Goal 4 = Quality education, Goal 12 = Responsible consumption and production			
4	Sustainable development goals, sports and physical activity: The localization of health-related sustainable development goals through sports in China: A narrative review	DAI J (2020)	*Risk Management and Healthcare Policy*	25	10.2147/RMHP.S257844
	Themes and keywords of the article	Built environment, elderly population, glocalization, healthy China, NCDs			
	SDGs mapped to this document	Goal 8 = Decent work and economic growth, Goal 15 = Life on land, Goal 17 = Partnerships for the goals			
5	What Does Innovation Mean to Nonprofit Practitioners? International Insights from Development and Peace-Building Nonprofits	SVENSSON PG (2020)	*Nonprofit and Voluntary Sector Quarterly*	24	10.1177/0899764019872009
	Themes and keywords of the article	Innovation, practitioner-driven, social change, social transformation, sport for development and peace (SDP), UN Sustainable Development Goals			
	SDGs mapped to this document	Goal 16 = Peace, justice and strong institutions, Goal 9 = Industry, innovation and infrastructure			

(*Continued*)

Table 2.1 (Continued)

Sr. no.	Title	Author	Journal	TC	DOI
6	Sport for development and peace and the environment: The case for policy, practice, and research	GIULIANOTTI R (2018)	*Sustainability* (Switzerland)	18	10.3390/su10072241
	Themes and keywords of the article Development, environment, peace, sport, sustainability, Sustainable Development Goals				
	SDGs mapped to this document Goal 8 = Decent work and economic growth, Goal 13 = Climate action, Goal 17 = Partnerships for the goals				
7	United nations sustainable development goals: Promoting health and well-being through physical education partnerships	LYNCH T (2016)	*Cogent Education*	16	10.1080/2331186X.2016.1188469
	Themes and keywords of the article Health, partnerships, physical education, strengths-based approach, Sustainable Development Goals, teacher education, well-being				
	SDGs mapped to this document Goal 4 = Quality education, Goal 12 = Responsible consumption and production				
8	The relationship between Walk Score® and perceived walkability in ultrahigh-density areas	KOOHSARI MJ (2021)	*Preventive Medicine Reports*	11	10.1016/j.pmedr.2021.101393
	Themes and keywords of the article Measurement, perceptions, sustainability, Sustainable Development Goals, urban design, walkable areas				
	SDGs mapped to this document Goal 3 = Good health and well-being, Goal 11 = Sustainable cities and communities				
9	Configuring relationships between state and non-state actors: a new conceptual approach for sport and development	LINDSEY I (2020)	*International Journal of Sport Policy and Politics*	11	10.1080/19406940.2019.1676812
	Themes and keywords of the article Civil society, government, non-governmental organizations (NGOs), partnership, sport for development and peace (SDP), sustainable				
	SDGs mapped to this document Goal 17 = Partnerships for the goals				
10	Debating the success of carbon-offsetting projects at sports mega-events. A case from the 2014 FIFA World Cup	CRABB LAH (2018)	*Journal of Sustainable Forestry*	9	10.1080/10549811.2017.1364652
	Themes and keywords of the article Carbon offsetting; Mato Grosso, reforestation, Sustainable Development Goals				
	SDGs mapped to this document Goal 13 = Climate action, Goal 15 = Life on land				

#	Article / Author (year) / Journal	TC	DOI
11	The distinct role of physical education in the context of Agenda 2030 and sustainable development goals: An explorative review and suggestions for future work — FRÃ–BERG A (2021) — *Sustainability (Switzerland)*	8	10.3390/su132111900
	Themes and keywords of the article — Agenda 2030, children and adolescents, physical education, school, sustainable development		
	SDGs mapped to this document — Goal 4 = Quality education, Goal 12 = Responsible consumption and production, Goal 15 = Life on land, Goal 17 = Partnerships for the goals		
12	Segmentation of passenger electric cars market in Poland — KUBICZEK J (2021) — *World Electric Vehicle Journal*	8	10.3390/wevj12010023
	Themes and keywords of the article — Classification methods, EV market, market segmentation, passenger electric cars		
	SDGs mapped to this document — Goal 7 = Affordable and clean energy, Goal 8 = Decent work and economic growth, Goal 17 = Partnerships for the goals		
13	Monitoring and evaluation of sport-based HIV/AIDS awareness programmes: Strengthening outcome indicators — MALEKA EN (2017) — *Sahara J*	7	10.1080/17290376.2016.1266506
	Themes and keywords of the article — HIV/AIDS, indicator, non-governmental organisations, outcome, performance assessment, sport-for-development		
	SDGs mapped to this document — Goal 3 = Good health and well-being, Goal 5 = Gender equality		
14	The role of dark commemorative and sport events in peaceful coexistence in the Western Balkans — A ULIGOJ M (2022) — *Journal of Sustainable Tourism*	6	10.1080/09669582.2021.1938090
	Themes and keywords of the article — Dark events, post-conflict, sports events, sustainable development goals, sustainable tourism, Western Balkans		
	SDGs mapped to this document — Goal 8 = Decent work and economic growth, Goal 12 = Responsible consumption and production		
15	Sport and sustainable development goals in Spain — CAMPILLO-SÁNCHEZ J (2021) — *Sustainability (Switzerland)*	6	10.3390/su13063505
	Themes and keywords of the article — Agenda 2030, physical activity, policy coherence, policy lever, sport, sustainable development goals		
	SDGs mapped to this document — Goal 15 = Life on land, Goal 17 = Partnerships for the goals		

Note: Table created from the Scopus database as of March 2023, where TC = total citation and DOI = digital object identifier

Dai and Menhas (2020) talked about the Chinese government's plan in achieving health-related SDGs by launching a "Healthy China 2030 plan" through sports and PA. The focus was on the localization of the health-related SDGs for the realization of SDGs at a global level. They suggested strong policies for a secure health system and that global health governance should be promoted. Svensson et al. (2020) conducted an in-depth interview with 49 "non-profit organization leaders of Sports and Development for Peace (SDP)". The findings of their study revealed the multifaceted meaning of non-profit innovation. In order to retain non-profit organizations, social transformation and continuous innovation and technical upgradation due to a lack of resources are needed. Giulianotti et al. (2018) stated that the announcement of SDGs lent increased interest to the problem of "environmental concerns and sustainability in the global development scenario" in the field of SDP. To achieve this sustainability agenda, the practice of SDP needs to be adopted on a day-to-day basis of our lives. Lynch (2016) mainly talked about PE in the context of SDG 3 and SDG 4, which addresses well-being and health and quality education respectively. They provided a lens on cross-sector "partnership" as an enabler for the implementation of SDGs. The United Nations recognized a solid gap between the information on partnership for all goals in action and a need to have to report on the ground level. Various teaching experiences in diverse roles were drawn upon when establishing community collaborations, these practices along with the reassurance and support provided by the international "outstanding" UK ITE programme research, which supported and provided the strength and drive for ongoing partnerships. The intention of the programme was to offer what was viewed as being "in the best interest of the children", which provided motivation. Overall, a global partnership can enable scope for creative contextual opportunities and facilitate the effective implementation of SDGs. Koohsari et al. (2021) are of the view that the built environment that surrounds us influences the levels of PA and is important for public health. They conducted a qualitative analysis of the Walk Score and perceived a walkable environment which consists of population density, public transport, recreational facilities, etc. They highlighted that the Walk Score correlated with several perceivable environment attributes in the context of ultrahigh-density areas in Asia.

Fröberg and Lundvall (2021) provided a heuristic tool so that scholars, policymakers, and practitioners could better grasp how various links may (or may not) increase the potential advantages of sports to development. For policymakers and practitioners, it made clear how crucial it is to move past the oversimplified concept of "partnership" that remains prevalent in the sports and development sectors. The relevance of various agreements of links between state and non-state actors involved in sports with respect to different SDGs and targets differs. Every nation is expected to set priorities and create implementation plans for SDGs that are compatible with their unique settings. Crabb (2018) is of the view that "[t]he United Nations Framework Convention on Climate Change" views using forests to mitigate climate change. Their inclusion in the SDGs has further demonstrated their significance in the sustainable development paradigm. Although the SDGs are intended to provide a worldwide framework for "equitable and sustainable development", their implementation poses new challenges for those who work in forestry. In order to evaluate the amount of project success, they employed an "ethnographic case study" to look at a "forestry-based carbon offsetting initiative in Mato Grosso, Brazil". The initiative gained international attention since it was developed to reduce emissions resulting from the construction of the "new football stadium designed for the 2014 FIFA World Cup in Mato Grosso". With the help of earlier work, this case study starts to problematize organizational and implementation challenges with carbon offsetting forestry operations. In order to demonstrate how "carbon offsetting initiatives that work towards a broad range of SDGs may fail, significant events in the project's implementation were used". However, there frequently still exists a gap among project goals, results, and participant experiences. The coordination of these three factors can be proven to be effective in influencing the success of such projects.

Fröberg and Lundvall (2021) raised a critical question as to how sustainability can be understood, framed, and integrated into PE in schools. They performed a review of the distinct role of PE in the context of Agenda 2030. The analysis was conducted on approximately 4,300 papers that were published between 2015 and 2021. Sports and PA have been recognized by numerous organizations and authorities as essential tools for achieving the SDGs. Despite this, it seems logical to distinguish the fields of PA and sports from PE because each field has its own distinctive policy and evidence chains. They concluded that not much research has been done in the area and that more attention is required from scholars and suggested adding an educational perspective to the field of SDGs and PE.

Maleka (2017) discusses the search for alternate forms of vehicle propulsion that aimed to advance SDGs and protect the environment, suggesting that it was driven by automobile manufacturers' strategies. Using electric motors in vehicles is an example of putting into action a sustainable agenda. The purpose of the paper was to show the market segmentation that had been carried out, with a "particular emphasis on finding the shared traits of the electric car models available on the Polish market and suggesting groupings for them". As one of the unusual studies in this area, the Polish EV market was segmented not only according to car size and class but also on the basis of performance and general vehicle quality. When the price was included as a second variable, the performed categorization remained unchanged. Additionally, it was suggested that the market be segmented into four categories: "Premium, City, Small, and Sport". Maleka (2017) aimed to enhance the options for employing indicators of "sport-based HIV/AIDS awareness programs of chosen South African NGOs" as a contribution to a general monitoring and evaluation framework. Seven staff from five chosen NGOs that integrate sports to offer HIV/AIDS programmes in South Africa took part in this qualitative study. Information on performance and indicators for sports-based HIV/AIDS awareness programmes was gathered using a variety of data collection tools, including "desktop review, narrative systematic review, document analysis, one-on-one interviews, and focus group interviews". This was the only study that related sports and health-related issues directly. This study provided a generic set of indicators to monitor the performance of sports-based HIV/AIDS awareness campaigns. They further suggested that cooperation from all the stakeholders at each level is essential for successful implementation and monitoring. Šuligoj and Kennell (2022), in order to analyse how "dark events affect post-conflict tourist development and peaceful cohabitation", used "sociological discourse analysis". This research clarified the relationship between "event tourism and SDG 16 focusing on peace, justice, and strong institutions" by integrating earlier analyses of "dark commemorative events" with an updated evaluation of sports events. This media-based analysis revealed similarities among "dark commemorative and sporting events' historical settings, dissonant heritage, and spectator's dark leisure habits". These troubling instances draw tourists from outside the region's boundaries as well as more scattered diasporas, worldwide media, and politicians. In order to encourage sustainable development, a lot of attention needs to be paid to the region's frequently occurring dark events, which are prominent for tourist inflows.

Campillo-Sánchez et al. (2021) stated sports as a valued tool for sustainable development. This is also recognized in the 2030 Agenda of SDGs itself and in the action plan for its successful implementation in Spain. In order to expand the scope of the SDGs, it is essential to carefully analyse both "potential synergies and current inconsistencies" that can strengthen or diminish the contribution of sports to sustainable development. It will be required to take into account the idea of "policy coherence" in both its vertical and horizontal dimensions, employing international suggestions in this respect as a guiding model. Local governments and other stakeholders must be involved in order to achieve the SDGs, supporting decision-making based on reliable and concrete common indicators.

Highly contributing countries

In order to understand which countries are contributing towards the research on sports and SDGs, the highly cited countries were analysed. It gives us an indication of awareness on how much focus on sports can help in achieving the SDG agenda in time. It is noteworthy to find that from the dataset with research contributions from 41 countries, only eight are highly cited when the threshold of minimum documents per country is kept at 5.

Table 2.2 Top eight highly cited countries

Sr. no.	Country	Documents	Citation
1	United Kingdom	16	127
2	Spain	13	107
3	United States	6	29
4	South Africa	6	14
5	China	5	32
6	Canada	5	31
7	Norway	5	27
8	Australia	5	21

Note: Table generated using VOSviewer software.

The above table and figure show the highest contribution of the United Kingdom with 16 documents and 127 citations.

Thematic evolution

The concept of SDGs became the research focus only after 2015, which is why if we look at the thematic evolution of research in this area in Figure 2.5, it indicates the time zone of research in this area as 2014 or 2015 onwards. The major thrust of research is especially seen in "sports for development and peace", which remained a highlight up to 2020. The three focus areas or emerging topics in the research remained the same up to 2023. A strong alignment of research on "Sports with the achievement of SDGs" could be seen in the current scenario. Furthermore, sports as a contrivance of achieving peace goals remained an all-time highlight in the whole decade in the dataset. The concept of sustainability is seen to be a highlight in the current time period also, giving an indication of future research scope in this area.

The thematic map of research in Figure 2.6 shows the four quadrants highlighting the themes of research. The first quadrant supports the niche areas "SDGs" and "social sustainability". It also highlights one more area, that is, "sports diplomacy",

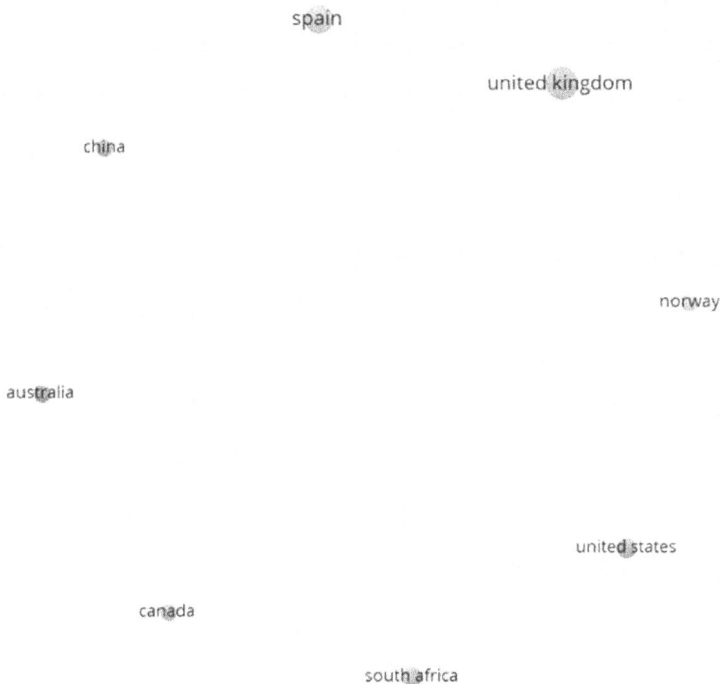

spain

united kingdom

china

norway

australia

united states

canada

south africa

Figure 2.4 Highly contributing countries.

Figure 2.5 Thematic evolution of the field of sports and SDG from 2014 to 2023.

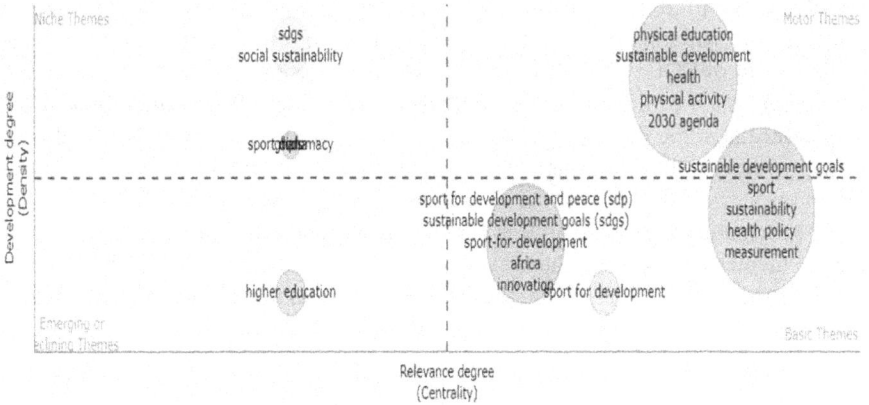

Figure 2.6 Thematic map of the research.

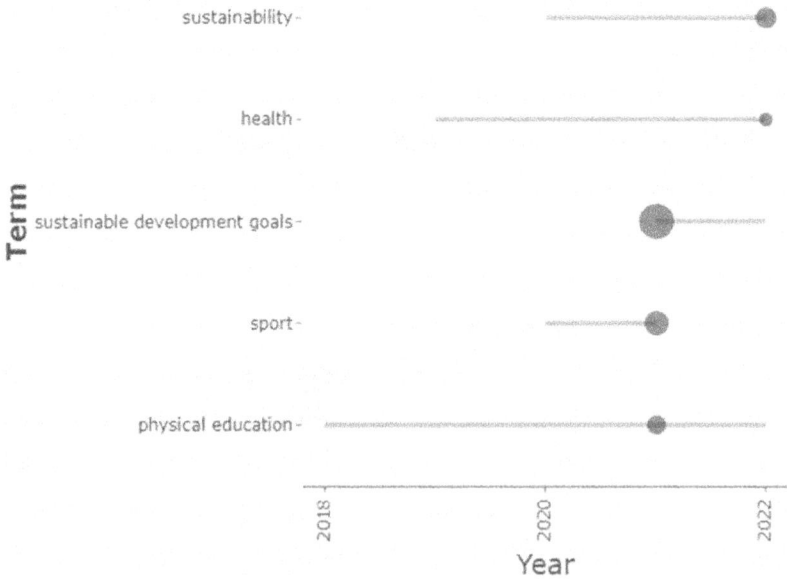

Figure 2.7 Most trending keywords.

to be one of the niche areas. Motor themes are the themes that have reached the peak of research, and many research studies are available in this area, especially on sports in relation to "health", "2030 Agenda", "physical activity", and "sustainable development". The basic themes of research in this area continue to be "sustainable development", "sports for development and peace", "sports in emerging nations", and innovations in the areas of sports. The area that is expected to gain more highlight in the near future as per this dataset is "sports for higher education".

Most trending topics highlighting the future research scope

The most trending keywords give us the idea about the areas that are being explored and that have future potential for research in the times to come. From 2018 to 2022, the dataset exhibits the most trending keywords to be "sustainability", "health", "sustainable development goals", "sports", and "physical education". It is worth mentioning that sports and PE in relation to health and SDGs are going to be the niche area of research in the coming years. It also highlights that if academicians are talking about sports and health as the important aspects, then they are going to be inculcated into "physical education in one way or another". It also draws certain policy implications for the governments to highlight the area of sports with the help of introducing certain norms in PE, starting from elementary to higher education.

Conclusion

The fundamental essence of sports is sustainability. The various dimensions of sports like team spirit, camaraderie, equality, partnerships, solidarity, heath, well-being, etc., point towards various sustainable goals. Sports have a great potential to achieve the greater dimensions of "Sustainable Development Goals". In various spheres of life, there is a strong need to develop infrastructure at various levels in order to promote sports in various economies. It will not only help achieve the developmental goals of the nations but also automatically help achieve the sustainable goals of the nations. PE and various policy implications at the grassroot level in order to promote sports are going to be the thrust areas of research in the near future.

References

Baena-Morales, S., Jerez-Mayorga, D., Delgado-Floody, P., & Martínez-Martínez, J. (2021). Sustainable development goals and physical education. A proposal for practice-based models. In *International Journal of Environmental Research and Public Health* (Vol. 18, Issue 4, pp. 1–18). MDPI. https://doi.org/10.3390/ijerph18042129

Campillo-Sánchez, J., Segarra-Vicens, E., Morales-Baños, V., & Díaz-Suárez, A. (2021). Sport and sustainable development goals in Spain. *Sustainability (Switzerland)*, 13(6). https://doi.org/10.3390/su13063505

Crabb, L. A. H. (2018). Debating the success of carbon-offsetting projects at sports mega-events. A case from the 2014 FIFA World Cup. *Journal of Sustainable Forestry*, 37(2), 178–196. https://doi.org/10.1080/10549811.2017.1364652

Dai, J., & Menhas, R. (2020). Sustainable development goals, sports and physical activity: The localization of health-related sustainable development goals through sports in China: A narrative review. *Risk Management and Healthcare Policy*, 13, 1419–1430. https://doi.org/10.2147/RMHP.S257844

Fröberg, A., & Lundvall, S. (2021). The distinct role of physical education in the context of Agenda 2030 and sustainable development goals: An explorative review and suggestions for future work. In *Sustainability (Switzerland)* (Vol. 13, Issue 21). MDPI. https://doi.org/10.3390/su132111900

Giulianotti, R., Darnell, S., Collison, H., & Howe, P. D. (2018). Sport for development and peace and the environment: The case for policy, practice, and research. *Sustainability (Switzerland)*, 10(7). https://doi.org/10.3390/su10072241

Jiménez-García, M., Ruiz-Chico, J., Peña-Sánchez, A. R., & López-Sánchez, J. A. (2020). A bibliometric analysis of sports tourism and sustainability (2002–2019). *Sustainability (Switzerland)*, 12(7). https://doi.org/10.3390/su12072840

Koohsari, M. J., McCormack, G. R., Shibata, A., Ishii, K., Yasunaga, A., Nakaya, T., & Oka, K. (2021). The relationship between Walk Score® and perceived walkability in ultrahigh density areas. *Preventive Medicine Reports*, 23. https://doi.org/10.1016/j.pmedr.2021.101393

Lindsey, I., & Darby, P. (2019). Sport and the sustainable development goals: Where is the policy coherence? *International Review for the Sociology of Sport*, 54(7), 793–812. https://doi.org/10.1177/1012690217752651

Lynch, T. (2016). United nations sustainable development goals: Promoting health and well-being through physical education partnerships. *Cogent Education*, 3(1). https://doi.org/10.1080/2331186X.2016.1188469

Maleka, E. N. (2017). Monitoring and evaluation of sport-based HIV/AIDS awareness programmes: Strengthening outcome indicators. *Sahara J*, 14(1), 1–21. https://doi.org/10.1080/17290376.2016.1266506

Šuligoj, M., & Kennell, J. (2022). The role of dark commemorative and sport events in peaceful coexistence in the Western Balkans. *Journal of Sustainable Tourism*, 30(2–3), 408–426. https://doi.org/10.1080/09669582.2021.1938090

Svensson, P. G., Mahoney, T. Q., & Hambrick, M. E. (2020). What does innovation mean to nonprofit practitioners? International insights from development and peace-building nonprofits. *Nonprofit and Voluntary Sector Quarterly*, 49(2), 380–398. https://doi.org/10.1177/0899764019872009

3 "Panem et circenses"

Investigating plate waste at major sport events

Emil Juvan and Miha Lesjak

Introduction

The 17 Sustainable Development Goals (SDGs) are an integral part of the 2030 Agenda for sustainable development, which serves as a *shared blueprint for peace and prosperity for people and the planet, now and into the future* (United Nations, 2015). Although well accepted among nations, it is clear that we as a society are not doing enough to fulfil the aim of the agenda. More specifically, a recent report shows that "*We must rise higher to rescue the Sustainable Development Goals*" (Guterres, 2023), clearly evidencing that the progress in reducing exploitation of natural and social resources is too slow. One of the key reasons for such slow progress may lie in the fact that the collected empirical data on the actual environmental and social impacts of human behaviour as well as drivers of such exploitative behaviour are insufficient. Understanding the actual burden may help raise awareness of consequences, and understanding the drivers increases our ability to change the harmful behaviour (Stern, 2005; Steg & Vlek, 2009).

Tourism contributes about 8% to climate change (Lenzen et al., 2018), and it substantially harms water resources and land (Gossling & Peeters, 2015). Food-related environmental impacts equate to about 43% of global land use, 60% of withdrawn fresh water, and about 23% of CO_2 emissions (Poore & Nemecek, 2018). Food waste is a particular issue in tourism as it represents a significant share of tourism's total waste (Juvan et al., 2023), up to 92% of food waste is avoidable (Betz et al., 2015; Papargyropoulou et al., 2016) and a large portion of food waste is landfilled (World Bank, 2018). When landfilled, 1 kg of food waste contributes 1.9 CO_2eq (BIO Intelligence, 2010) and produces methane, which further burdens ozone and by so doing contributes to climate change. Knowledge on food consumption lags behind our understanding of other aspects of tourisms' sustainability (Gossling & Peeters, 2015), but the availability of empirical evidence is growing (e.g. Juvan et al., 2018; Leverenz et al., 2020; Wang et al., 2021). Plate waste is uneaten food left behind by tourists, when they serve but do not eat all the food they self-served. Tourists leave behind about 30% of their food uneaten (Papargyropoulou et al., 2016) and report a number of reasons for not finishing the food they self-served (Dolnicar & Juvan, 2019). Understanding plate waste is particularly interesting as it targets waste produced without any reasonable need for wasting food. More specifically, self-serving

DOI: 10.4324/9781003384786-3

plate waste refers to the food that was served by the tourists but left uneaten on the plate. So the tourist was in full control of choosing the type of food, the amount of food, the food combination, and other aspects of the portion.

The aim of this chapter is to highlight the actual plate waste generated at a large sport event venue by different types of spectators. By so doing, we extend our understanding of the volume of plate waste generated by different types of tourists, specific to the context of mega sport events. A key advantage of this chapter is that it provides the objectively measured volume of plate waste, which ensures high validity of data on plate waste volume. In addition, we empirically validate theoretically supported drivers of plate waste. These findings directly contribute to SDG 12.3 because they inform the amount of food waste and approaches to reduce global food waste at consumer levels.

Literature review

Major sport events have major economic, social, and environmental consequences (Dwyer, 2015); thus, they affect achieving global SDGs. Sport events and festivals play an important role in tourism economy (Page & Connell, 2020; Wise et al., 2021a), rapidly expanding the segment of the leisure travel market (Getz & Page, 2016), and are drivers of significant economic, socio-cultural, and environmental impacts on host destinations (Dwyer, 2015). Typically, international sports events improve the competitiveness of host destinations (Wise, 2020; Dimanche, 2003; Chalip & Fairley, 2019), increase visitor numbers (Li & Jago, 2013; Bazzanella et al., 2023), and improve destination tourism revenues (Fourie & Santana-Gallego, 2011; Lesjak et al., 2016). The organization of sports events brings greater media attention and, consequently, greater visibility and promotion of the destination.

Sustainability is at the forefront of current discussions on sports tourism and has become an important topic not only for organizers but also for spectators and local residents of host areas. Defining sustainable and responsible events requires considering that events have to be *"sensitive to the economic, sociocultural and environmental needs within the local host community, and organized in such a way as to optimize the net holistic (positive) output"* (Raj & Musgrave, 2009, p. 25). Several authors studied sustainable management of events. More specifically, Hede (2007) studied triple bottom line (TBL) in special event management and concludes that the underlying principles of TBL should be applied already at the planning stage of event management. Additionally, the study demonstrates a heterogeneity in stakeholder support for sustainable orientation of events. It is often the case that event planners have limited scope and knowledge of sustainable events, including the knowledge about sustainability programmes. This leads to slow progress in increasing the sustainability of events (Park & Boo, 2010). Yet, the evidence of sustainable progress is visible in the context of small-scale sports events where community partnership in events organization leads to improved sustainability of events (Gibson et al., 2012).

The environmental consequences of major sports events have become a crucial driver of long-term costs and benefits of events (Getz, 2012). Therefore, the implementation of sustainable principles within major sports events *"must be inte-grated, planned and should reflect the elements of quality management frameworks"* (Musgrave, 2011, p. 269). Organizing sustainability-oriented events should follow sustainable measures and solutions, allowing for documentation of environmen-tal impacts. More specifically, sports event organizers should be able to measure and report waste management practices, use of soft mobility, use of energy-saving technology, water use, etc. (Chalip, 2014; Collins et al., 2009). Effective sustainable management should follow the notion that drivers of pro-environmental behaviour differ across different types of behaviour (Juvan & Dolnicar, 2017), that environmen-tal impacts or consequences differ across different contexts of tourism (Juvan et al., 2023), and that sports event spectators are a heterogeneous market with potentially different dispositions for pro-sustainable behaviours (Dolnicar & Grün, 2009).

CO_2 emissions are the most concerning environmental issue of tourism because (1) emissions substantially contribute to climate change, (2) available data on con-tribution are rather reserved, and (3) tourism aviation emissions are growing faster than expected (Gössling et al., 2023; Lenzen et al., 2018). We must find ways to reduce tourism CO_2 emissions, and air transportation appears as the most often tar-geted context of tourism where reductions must occur. Air transport emissions are the major driver of tourism's contribution to climate change (Gossling et al., 2023; Gössling & Humpe, 2020), thus also a highly attractive context of increasing the sustainability of tourism. However, the high dependence of international tourism on flying also makes air transport-related reductions extremely difficult to achieve. On the contrary, wasting food is not needed at all, and tourism food waste-related emissions account for about 87% of the road transport emissions (FAO, 2019), making this context of environmental impacts of tourism highly attractive for increasing tourism's sustainability. More specifically, international tourism heav-ily depends on flying; thus, substantial reduction of flying would have substantial negative economic consequences on many destinations (e.g. island destinations). On the other hand, tourists have no objective reason to waste food, and reducing food waste can substantially reduce the negative impacts on climate and biodi-versity (Beretta & Hellweg, 2019). Empirical evidence shows that tourists waste between 14 and 400 grams of served food (Juvan et al., 2018; Papargyropoulou et al., 2016). Although the food waste volume is different across hospitality settings and countries, it is agreed that in tourism about one-third of food served is wasted.

Not wasting food can be considered as private sphere environmentally signifi-cant behaviour (Stern, 2000). Such behaviour is driven by a number of factors, of which personality and contextual factors may be considered as prevailing. More specifically, attitudinal factors such as the belief that wasting food harms the envi-ronment (awareness of consequences) or holding oneself responsible for reduc-ing food waste (personal norm) may lead people to eat up all they serve on their plate. Nevertheless, in the hospitality and tourism context, tourists waste food also because they are unfamiliar with the food served or how to combine food available at food buffets (e.g. knowledge) or because of the food service style (e.g. served by the kitchen staff of self-serving buffets) (Juvan & Dolnicar, 2019). Perceived food variety and abundance may also serve as the driver of food waste (Juvan et al.,

2017). Other empirically tested drivers of food waste in the tourism context include habits, food offering, values, and norms.

Although conclusions on the importance of environmental impacts depends on the assessor (Chersulich et al., 2020), environmental impacts of sports events should be treated with particular attention. Major sports events bring together different groups of tourists who travel for various reasons and behave differently, thus generating different environmental footprints (Lesjak et al., 2017), depending on different types of behaviour and driven by different causes. For example, in the context of food-related behaviour, (1) food is being served and handled at venues (e.g. under temporary infrastructures such as tents) different from that of the mainstream tourism, (2) events bring together international and domestic spectators with different eating and nutritional requirements, and (3) large sports events need to provide food for a large volume of people, much higher than any typical tourism venue. Given that situational characteristics play a significant role in sustainable behaviour (Stern, 2000), we can reasonably assume that behaviours and impacts of such different behaviours vary across different sports events and most likely across different groups of attendees at one sports event.

This literature review allows the conclusion that food waste is a major driver of environmental costs of tourism, including sports tourism, which involves sports events. Furthermore, food volume varies across different contexts and depends on different drivers. Empirical assessment of the food waste volume and potential drivers in the context of sports events can serve as a starting point for (1) acknowledging the extent of the food waste problem at sports events and (2) the development of context-specific empirically supported interventions for managing food waste at sports events.

Methodology

The study was conducted at the FIS Ski Jumping World Cup Finals, a major winter sports event in Slovenia. The sports event took place between 24 and 27 March 2022 and was attended by 59,000 spectators. The study employed a combination of observational and survey research methods. Plate waste was observed and measured by research assistants at four different locations serving food to four different types of event attendees. Plates were de-served by the serving staff and brought to the kitchen areas, where research assistants supervised the disposal process. The serving staff discarded plate waste to a designated food waste bin, and the research assistants weighted the discarded plate waste. Another group of research assistants counted the number of guests entering the food service area, observed the average time guests spent in the VIP area, and invited guests to fill in the survey upon exiting the VIP area. Research assistants received a two-hour training on measuring plate waste, guest observation, and distribution of survey. They also received a tour of the VIP premises to familiarize themselves with the working process.

This study reports on plate waste at three different VIP areas, each for a specific type of spectators. More specifically, we provide the actual plate waste for visitors at the regular VIP area, visitors at the upscale VIP area, and visitors (journalists and reporters) at the media VIP area. Drivers of plate waste are reported only for regular and upscale VIP visitors because we were not able to intercept visitors of the media VIP area.

We calculated average plate waste per person as the total plate waste per day divided by the number of eaters for each respective VIP area. Research assistants observed the average time spent in regular and upscale VIP areas and made notes in their observation diary. Maintaining discretion was crucial therefore, and no other formal way of recording the time spent or other aspects of behaviour in the VIP areas was possible. Research assistants were asked to report the time spent in the VIP areas as a share of attendees whose time was short (up to 30 minutes), medium (between 30 and 60 minutes), and long (more than one hour).

We measured potential drivers of plate waste by a means of an online survey, only for regular and upscale VIP visitors. Upon leaving the VIP area, research assistants invited guests to provide feedback on the VIP experience using an online survey. Visitors were explained that the survey takes about three minutes and is accessible using a QR code displayed by the research assistants. Details about the food waste-related content of the survey were withheld from respondents so as to minimize respondents' social desirability bias (Juvan & Dolnicar, 2016) and to reduce respondents declining to participate in the survey. Study participants scanned the QR code to access the survey and filled in the survey on their own, after leaving the VIP area. The survey involved questions about food served, service, and awareness of food waste and its environmental costs. More specifically, the respondents rated volume of food available (0: very little, 10: a lot), diversity of food (0: not diverse at all, 10: very diverse), and quality of food (0: very low quality, 10: very high quality). Next, we asked about the experiences related to service. More specifically, the respondents rated the time needed to get to the food (0: very little time, 10: a lot of time), staff friendliness (0: not friendly at all, 10: very friendly), and enjoyment of the food (0: not at all, 10: very much enjoyed). The study participants were also asked to report the perceived amount of food they left behind uneaten in the plate (0: no food left uneaten, 10: a lot of food left uneaten) and how many times they attended the food buffet (1: only once, 2: twice, and 3: three or more times). The last section of the survey measured beliefs about food waste awareness and its ecological impacts, more specifically awareness of negative environmental consequences of food waste (0: very little, 10: very high), ascription of responsibility for reducing food waste (0: very little, 10: very high), and moral obligation for reducing food waste (0: not at all, 10: a lot). Given the normal distribution of data, we used *t*-test statistics for detecting the association between drivers of plate waste and reported plate waste.

The food menu is different across different VIP areas, reflecting the overall price paid for entering the VIP. The "regular VIP" menu consisted of five different food items (sausage, beef stew, beans stew, barley soup, and polenta). The "upscale VIP" guests were offered 17 different dishes (e.g. beef stew, sausages, pork roast, cabbage and turnips, polenta, savoury cheese rolls, cold meat cuts, and different types of pastas, salads, and desserts). Service style (self-serving buffet), atmosphere, and other food service context-dependent characteristics were the same. The price for entering was different, with the price for upscale VIPs being twice higher than the price for regular VIP.

The study population were guests of different VIP sections with different socio-economic backgrounds. While measuring their actual socio-demographic background was not possible, the way VIP guests obtained the ticket to VIP serves as a proxy for their socio-demographic background. Most of the "regular VIP" guests received the pass from the organizers of the major sporting event, and majority were local entrepreneurs and other local public personas. The "upscale VIP" guests obtained their pass via major sponsors and were major nationally important entrepreneurs, businessmen, and other influential individuals.

Results

A total of 270 guests of regular and upscale VIPs responded to the survey invitation, of which 52% were male respondents. The average age of the respondents was 42 years, 3% of the participants were aged between 14 and 17 years, 21% were between 19 and 30 years of age, and 50% were between 31 and 50 years of age. Nearly 53% of the respondents were return visitors to VIP areas at this particular event, and the rest visited the VIP areas at this event for the first time. Analysis of respondents' behaviour at the VIP areas shows that significantly more upscale VIP than regular VIP visitors visited the food buffet more than one time ($X_{(2)} = 9,697$; $p = .008$; phi = .118). Observations of the average time spent at the VIP areas show that substantially (about 70%) more observed visitors at the upscale VIP area spent more than one hour at the VIP area. Only about 10% of visitors of the regular VIP area stayed for one hour or more.

Table 3.1 provides measurements of plate waste for four different food-serving locations at the sports event. Data suggest that the average plate waste increases with the quality of the VIP. More specifically, the average plate waste per person in media (journalist) VIPs is five times lower than the average plate waste in a regular VIP and 12 times lower than the plate waste in an upscale VIP. The average plate waste per person in upscale VIPs is two times higher than the plate waste in a regular VIP.

Food service style was the same across all three locations, but the menus and clientele were different. More specifically, the clientele differed in how they obtained the VIP pass, which serves also as a proxy for their material and social status.

Table 3.1 Food waste for three different types of food-serving locations

VIP	Day 1		Day 2		Day 3		Avg. total (g)
	Eaters	Avg. waste (g)	Eaters	Avg. waste (g)	Eaters	Avg. waste (g)	
Regular	200	47.5	900	59.5	900	72.6	59.8
Upscale	360	114	990	132	870	155	133.6
Media	250	12	250	9	250	12	11

Therefore, we analysed the differences in personal characteristics of visitors, which can serve as drivers of plate waste, at regular and upscale VIP areas. The most important finding is that no significant differences between the visitors of two different VIP areas were identified for awareness of consequences, ascription of responsibility, and personal norms. In addition, no differences were identified also for the perceived amount of plate waste between the visitors of the two types of VIPs. More specifically, respondents from both types of VIP report leaving behind almost no uneaten food (VIP-regular, M = 1.89; VIP-upscale, M = 2.04).

Table 3.2 shows the results of *t*-test statistics on the food and dining experiences as well as attitudinal factors of visitors at the two VIP areas. It can be seen that visitors were quite positive about the food (e.g. amount, quality, and time needed to get the food). They also reported quite positively about familiarization with the food and enjoyment of the food. Nevertheless, statistical differences in two potential drivers of food waste were identified. Visitors of the upscale VIP report a significantly higher level (M = 7.25) of perceived food variability than their counterparts from the regular VIP (M = 6.67; *p* = .014). Visitors of the upscale VIP are also significantly more favourable (M = 8.24) about staff friendliness than their counterparts from the regular VIP (M = 7.58, *p* = .001).

Table 3.2 Potential drivers of food waste for regular and upscale VIPs

		N	*Mean*	*Std. deviation*	*Std. error mean*
Contextual factors					
How much food was available?	VIP-regular	154	7.19	2.335	.188
	VIP-upscale	125	7.32	2.302	.206
How many different dishes were available?	**VIP-regular**	**154**	**6.67**	**2.264**	**.182**
	VIP-upscale	**119**	**7.25**	**2.222**	**.204**
What was the quality of the food?	VIP-regular	153	7.08	2.212	.179
	VIP-upscale	123	7.28	2.278	.205
How familiar were you with the food available?	VIP-regular	155	7.66	2.405	.193
	VIP-upscale	124	7.90	2.197	.197
How much time did you spend to serve the food?	VIP-regular	153	2.63	2.585	.209
	VIP-upscale	122	2.59	2.625	.238
How friendly was the staff?	**VIP-regular**	**153**	**7.58**	**2.465**	**.199**
	VIP-upscale	**119**	**8.24**	**2.221**	**.204**
How much did you enjoy the food?	VIP-regular	151	7.23	2.381	.194
	VIP-upscale	119	7.61	2.267	.208
Attitudinal factors					
How much food did you leave behind uneaten?	VIP-regular	152	1.98	2.332	.189
	VIP-upscale	119	2.04	2.482	.228
What is the environmental threat of food waste (AC)?	VIP-regular	150	7.25	2.572	.210
	VIP-upscale	115	6.93	2.571	.240
How much you can personally do to reduce food waste (AR)?	VIP-regular	150	7.17	2.401	.196
	VIP-upscale	116	6.92	2.547	.237
How morally obliged do you feel for reducing food waste (PN)?	VIP-regular	149	7.36	2.436	.200
	VIP-upscale	114	7.39	2.276	.213

Table 3.3 Perceived amount of food left on the plate uneaten by extreme groups based on key pro-environmental beliefs about food waste

	N	Mean perceived amount of uneaten food (0: no food, 10: a lot of food)		Statistics
		LOW	HIGH	
Awareness of consequences (AC)	259	3.46	1.12	$t = 8.348_{(186)}, p = .001$
Ascription of responsibility (AR)	264	3.57	.48	$t = 10.458_{(179)}, p = .001$
Moral obligation (PN)	261	3.56	1.06	$t = 9.172_{(195)}, p = .001$

An important finding is also that no differences exist in the number of visitors with different levels of pro-environmental beliefs (attitudinal factors) across the two VIPs. This means that the two VIPs had similar shares of visitors with similar levels of pro-environmental beliefs.

Next (Table 3.3), we compared the perceived amount of uneaten food between two extreme groups of respondents according to the key attitudinal factors of environmentally significant behaviour. Results showed a significant association between all three beliefs and perceived food waste. More specifically, respondents with lower levels of awareness of consequences, ascription of responsibility, and personal norms reported higher levels of uneaten food (plate waste).

Looking at the serving behaviour, that is, how many times attendees visited the food buffet, we see that a significantly higher share (72%) of VIP-upscale attendees and then VIP-regular attendees (54%) reported visiting the buffet more than one time ($X_2 = 8.322_{(1)}, p = .004$). Respondents who visited the buffet twice reported leaving significantly less food behind than those visiting the buffet once only ($F_{(2,266)} = 3.154, p = 0.44$). This suggests the occurrence of overserving (Dolnicar & Juvan, 2019), which may be driven by the VIP context. More specifically, those attending a buffet only once may overserve their plates when visiting the buffet. Those attending a buffet more than once may be more conscious in how much food they serve (Kallbekken & Sælen, 2013). However, researchers observing VIP attendees reported that a significantly higher share of VIP-upscale visitors stayed 1 hour or longer than visitors of VIP-regular. The amount of time spent at the dining area (such as VIP) may drive plate waste (Dolnicar & Juvan, 2019).

Discussion and conclusions

To the best of the authors' knowledge, the present chapter presents, for the first time, empirical evidence on the volume of plate waste and potential drivers of plate waste among different types of spectators in the context of large sports events. Guests at regular VIPs leave about 60 g of food behind, and guests at the upscale VIP area leave about three times as much. Comparing the average plate waste of the two groups of respondents with the data available on plate waste from other contexts of tourism (e.g. Leverenz et al., 2020; Juvan et al., 2017), one can observe substantially higher plate waste generated by respondents in this research. More specifically, hotel guests in the same country (Juvan et al., 2017) produce about 16 g of plate waste at breakfast and about 50 g of plate waste at dinner. Catering events in Germany (similar to the context of this study) produce between 74 g and 280 g of buffet food leftovers, which includes also food left behind at the buffet and not only food left behind on guest plates. This study shows 134 g of plate waste only.

However, guests at VIPs of a higher quality, with more food on offer, leave substantially more plate waste (on average 134 g). Journalists, on the other hand, leave on average 11 g of plate waste. These results show that food service context and purpose of attending events may be important drivers of plate waste. More specifically, VIP guests visited food service areas as part of their enjoyment of the event; they also reported a significantly higher level of food diversity offered and a higher level of perceived staff friendliness. The two are attributes of a more enjoyable atmosphere, which is a key attribute of events addressing hedonic drivers of attendance (Getz & Page, 2016). On the other hand, journalists visit events as part of their job and attend food service areas to satisfy their hunger. An important finding is also that other food offering attributes (e.g. food quality, familiarization, and time spent for serving) hypothesized to increase plate waste (Dolnicar & Juvan, 2019) were perceived comparatively well yet not significantly differently across the VIPs. This suggests that in the context of this event, food variety and service atmosphere (staff friendliness) played a significant role as drivers of food plate waste.

Following the theory of environmentally significant behaviour (Stern, 2000), this study corroborates the association between key attitudinal drivers of environmentally significant behaviour and self-reported plate waste. It is evident that individuals with higher levels of reported pro-environmental attitudinal variables report lower levels of plate waste. However, caution is needed when interpreting this association because it may exist due to social desirability bias or cognitive dissonance (Juvan & Dolnicar, 2014, 2016). During the survey, respondents' privacy was ensured; thus, the impact of social desirability bias may be low (Krosnick, 1999). In addition, low perception of plate waste among respondents with high levels of pro-environmental beliefs may be driven by the mechanism of preventing cognitive dissonance – a well-known mechanism for managing the attitude–behaviour gap (Juvan & Dolnicar, 2021).

This chapter adds important empirical evidence to the existing body of knowledge on the extent and potential drivers of plate waste in the tourism sector. Evidence collected at a specific major winter sports event show that event attendees (spectators) produce a substantially higher amount of plate waste than their counterparts in other contexts of tourism. Furthermore, we corroborate existing evidence on the role of attitudinal and contextual factors affecting plate waste in the tourism context. More specifically, awareness of consequences, ascription of responsibility, and personal norms are negatively associated with reported levels of plate waste, and food variety appears to lead to more visits to food stations for tasting, which eventually leads to more plate waste, due to either overserving or disliking the food (Juvan et al., 2017).

We must reduce the amount of food waste in tourism, and this chapter indicates that the amount of plate waste in the context of sports tourism (among spectators) is substantially higher than that in other contexts of tourism. Sports event organizers must put in place interventions that will address attitudinal and contextual drivers of plate waste. Such interventions must not only increase pro-environmental beliefs, likely less effective for reducing plate waste, but also re-design existing food service defaults, for example, reducing the perceived food variety or limiting the maximum amount spent in the VIP area. Another promising intervention could be rewarding (e.g. Dolnicar et al., 2019) responsible food behaviour with discounts on VIP tickets for next events or providing tasting stations where visitors can try all varieties of food, without overserving.

A key limitation of the present study is the lack of guest mix data of VIP visitors. This would allow for an extensive analysis of the associations between visitors' characteristics and their food waste-related behaviour. Collecting this kind of data was impossible for the current study, so the authors relied on observational data and used proxies, such as social and material status related to the way specific visitors obtain their VIP pass. Future studies should look at plate waste at other types of sport events and other forms of waste, for example, buffet waste that requires developing interventions targeted at VIP staff, rather than VIP guests.

References

Bazzanella, F., Schnitzer, M., Peters, M., & Fabian B. B. (2023). The role of sports events in developing tourism destinations: A systematized review and future research agenda. *Journal of Sport & Tourism*. https://doi.org/10.1080/14775085.2023.2186925

Beretta, C., & Hellweg, S. (2019). Potential environmental benefits from food waste prevention in the food service sector. *Resources, Conservation & Recycling*, 147, 169–178.

Betz, A., Buchli, J., GVobel, C., & Mueller, C. (2015). Food waste in the Swiss food service industry – magnitude and potential for reduction. *Waste Management*, 35, 218–226.

BIO Intelligence (2010). Preparatory study on food waste across EU27. Retrieved April 26 2023 online https://ec.europa.eu/environment/eussd/pdf/bio_foodwaste_report.pdf

Chalip, L. (2014). From legacy to leverage. In J. Grix (Ed.), *Leveraging legacies from sports mega-events* (pp. 2–12). Basingstoke: Palgrave Macmillan.

Chalip. L., & Fairley, S. (2019). Thinking strategically about sport events. *Journal of Sport & Tourism*, 23(4), 155–158.

Collins, A., Jones, C., & Munday, M. (2009). Assessing the environmental impacts of mega sporting events: Two options? *Tourism Management*, 30(6), 828–837.

Dimanche, F. (2003). *The role of sport events in destination marketing*. Paper presented at the AIEST 53rd Congress in Sport and Tourism, Athens, Greece.

Dolnicar, S., & Grün, B. (2009). Environmentally friendly behavior: Can heterogeneity among individuals and contexts/environments be harvested for improved sustainable management? *Environment and Behavior*, 41(5), 693–714.

Dwyer, L. (2015). Triple bottom line reporting as a basis for sustainable tourism: Opportunities and challenges. *Acta Turistica*, 27, 33–62.

FAO. (2010). *Food wastage footprint & climate change*. Retrieved April 26, 2023 from www.fao.org/fileadmin/templates/nr/sustainability_pathways/docs/FWF_and_climate_change.pdf2019

FAO. (2019). The state of food and agriculture report: Moving forward on food loss and waste reduction. *Food and Agriculture Organization of the United Nations*. Retrieved from www.fao.org/3/ca6030en/ca6030en.pdf

Fourie, J., & Santana-Gallego, M. (2011). The impact of mega-sport events on tourist arrivals. *Tourism Management*, 32(6), 1364–1370.

Getz, D. (2012). Event studies: Discourses and future directions. *Event Management*, 16(2), 171–187.

Getz, D., & Page, S. J. (2016). Progress and prospects for event tourism research. *Tourism Management*, 52, 593–631.

Gibson, H. J., Kaplanidou, K., & Kang, S. J. (2012). Small-scale event sport tourism: A case study in sustainable tourism. *Sport Management Review*, 15(2), 160–170.

Gössling, S., Balas, M., Mayer, & Sun, Y-Y. (2023). A review of tourism and climate change mitigation: The scales, scopes, stakeholders and strategies of carbon management. *Tourism Management*, 104681.

Gössling, S., & Humpe, A. (2020). The global scale, distribution and growth of aviation: Implications for climate change. *Global Environmental Change*, 65, 102194.

Gössling, S., & Peeters, P. (2015). Assessing tourism's global environmental impact 1900–2050. *Journal of Sustainable Tourism*, 23(5), 639–659.

Guterres, A. (2023). *The sustainable development goals report 2022*. Retrieved April 26, 2023 from https://unstats.un.org/sdgs/report/2022/

Hede, A. M. (2007). Managing special events in the new era of the triple bottom line. *Event Management*, 11(½), 13–22.

Juvan, E., & Dolnicar, S. (2014). The attitude–behaviour gap in sustainable tourism. *Annals of Tourism Research*, 48, 76–95.

Juvan, E., & Dolnicar, S. (2016). Measuring environmentally sustainable tourist behaviour. *Annals of Tourism Research*, 59, 30–44.

Juvan, E., & Dolnicar, S. (2019). Drivers of plate waste: A mini theory of action based on staff observations. *Annals of Tourism Research*, 78, 102731.

Juvan, E., & Dolnicar, S. (2021). The excuses tourists use to justify environmentally unfriendly behaviours. *Tourism Management*, 83, 104253.

Juvan, E., Grün, B., & Dolnicar, S. (2018). Biting off more than they can chew: Food waste at hotel breakfast buffets. *Journal of Travel Research*, 57(2), 232–242.

Juvan, E., Grün, B., & Dolnicar, S. (2023). Waste production patterns in hotels and restaurants: An intra-sectoral segmentation approach. *Annals of Tourism Research Empirical Insights*, 4, 100090.

Kallbekken, S., & Sælen, H. (2013). Nudging' hotel guests to reduce food waste as a win–win environmental measure. *Economics Letters*, 119(3), 325–327.

Krosnick, J. A. (1999). Survey research. *Annual Review of Psychology*, 50, 537–567.

Lenzen, M., Sun, Y. Y., Faturay, F., Ting, Y. P., Geschke, A., & Malik, A. (2018). The carbon footprint of global tourism. *Nature Climate Change*, 8(6), 522–528.

Lesjak, M., Axelsson, E., & Mekinc, J. (2017). Sports spectators tourism reason when attending major sporting events: Euro Basket 2013, Koper, Slovenia. *European Journal of Tourism Research*, 16, 74–91.

Li, S., & Jago, L. (2013). Evaluating economic impacts of major sports events – a meta analysis of the key trends. *Current Issues in Tourism*, 16(6), 591–611. https://doi.org/10.1080/13683500.2012.736482.

Musgrave, J. (2011). Moving towards responsible events management. *Worldwide Hospitality and Tourism Themes*, 3(3), 258–274.

Page, S. J., & Connell, J. (2020). *Tourism: A modern synthesis*. London: Routledge.

Papargyropoulou, E., Wright, N., Lozano, R., Steinberger, J., Padfield, R., & Ujang, Z. (2016). Conceptual framework for the study of food waste generation and prevention in the hospitality sector. *Waste Management*, 49, 326–336.

Park, E., & Boo, S. (2010). An assessment of convention tourism's potential contribution to environmentally sustainable growth. *Journal of Sustainable Tourism*, 18(1), 95–113.

Poore, J., & Nemecek, T. (2018). Reducing food's environmental impacts through producers and consumers. *Science*, 360(6392), 987–992.

Raj, R., & Musgrave, J. (Eds.) (2009). *Event management and sustainability*. London: CABI.

Steg, L., & Vlek, C. (2009). Encouraging pro-environmental behaviour: An integrative review and research agenda. *Journal of Environmental Psychology*, 29, 309–317.

Stern, P. C. (2000). Toward a coherent theory of environmentally significant behavior. *Journal of Social Issues*, 56(3), 407–424.

Stern, P. C. (2005). Understanding individuals' environmentally significant behavior. *Environmental Law Reporter News and Analysis*, 35, 10785.

Tomino, C. A., Perić, M., & Wise, N. (2020). Assessing and considering the wider impacts of sport-tourism events: A research agenda review of sustainability and strategic planning elements. *Sustainability*, 12(11), 4473.

United Nations. (2015). *The 17 goals*. Retrieved on April 26, 2023 from https://sdgs.un.org/goals

Wang, L., Filimonau, V., & Le, Y. (2021). Exploring the patterns of food waste generation by tourists in a popular destination. *Journal of Cleaner Production*, 279, 123890.

Wise, N. (2020). Eventful futures and triple bottom line impacts: BRICS, image regeneration and competitiveness. *Journal of Place Management and Development*, 13(1), 89–100. https://doi.org/10.1108/JPMD-10-2019-0087

Wise, N., Đurkin Badurina, J., & Perić, M. (2021a). Assessing residents' perceptions of urban placemaking prior to hosting a major cultural event. *International Journal of Event and Festival Management*, 12(1), 51–69.

World Bank. (2018). *What a waste 2.0: A global snapshot of solid waste management to 2050*. Retrieved on February 21, 2023, from World Bank www.worldbank.org/en/news/infographic/2018/09/20/what-a-waste-20-a-global-snapshot-of-solid-waste-management-to-2050

4 Sport events and travel carbon footprint

Seeking an optimal balance?

Hrvoje Grofelnik, Marko Perić, and Nicholas Wise

Introduction

Outdoor sporting events are the core of sports tourism (Deery et al., 2004; Perić et al., 2021). Focusing on this unique connection between people, places, and activities (Weed & Bull, 2009; Chersulich & Perić, 2022), sports tourism has become a global business for many stakeholders (Pedauga et al., 2022; Weed, 2020). Undoubtedly, events, and more specifically sporting events, can foster new economic and social development for hosting destinations (Wise & Maguire, 2022). This implies increased visibility in media, new employment opportunities, tourism growth, increased revenues, and increased social capital (Ritchie et al., 2020; Wallstam & Kronenberg, 2022; Zhang & Park, 2015; Zhou et al., 2021). However, one cannot neglect the negative sides of sports events. These can lead to higher prices for products and housing, security and cultural problems, and increased pollution (Ahmed, 2017; Liu et al., 2017; Polcsik & Perényi, 2022).

The latter is particularly significant for tourism in general and sports tourism in particular. However, waste, noise, and air pollution at venues and within local host communities are recognized as major problems (Perić, 2018; Polcsik & Perényi, 2022; Zhang et al., 2022). Moreover, pollution and carbon generated by travelling to and from sporting events remain a wide open area of research (Dolf & Teehan, 2015; Wicker, 2019; Pereira et al., 2019; Grofelnik et al., 2020; Cooper & McCullough, 2021; Thormann et al., 2022). As in any other type of traditional tourism (i.e. with the exemption of virtual or e-tourism), both competitors and spectators need to travel to reach the event sites. Unless they travel on foot or by bike, travelling to and from destinations generates carbon dioxide (CO_2) emissions. These emissions are one of the largest contributors to greenhouse effects and global warming (Lenzen et al., 2018; Van Fan et al., 2018).

These concerns are directly related to the United Nations Sustainable Development Goals (SDGs). Goals 12 (Responsible consumption and production) and 13 (Climate action) are specifically concerned with sustainable consumption and product patterns, as well as reducing the impact of climate change. However, the amount of CO_2 emissions depends primarily on the travel distance and type, and capacity and efficiency of the transport mode that is used. Also, travel style has a significant impact on personal carbon emissions. In general, using public transport

DOI: 10.4324/9781003384786-4

and sharing transport result in lower personal carbon footprints and are more eco-logically acceptable (Sinha et al., 2019; Haase, 2022). Complete elimination of CO_2 emissions from travelling is far-fetched, but it is important to bring to atten-tion strategies on how to reduce participants' travel CO_2 emissions. Therefore, this research offers managerial implications to help achieve sustainable development in a distinct way.

Recent travel- and mobility-related research has considered the impact of such work in relation to the SDGs (Boluk et al., 2019; Moyle et al., 2022; Nunkoo et al., 2021; Rasoolimanesh et al., 2020). A challenge with such research is that most studies consider how research is aligning with SDGs, but only few studies are seeking to put forward managerial implications on how to meet these challenges (Fodness, 2017; Hall, 2019). With many papers theorizing tourism and mobility in relation to SDGs, more studies need empirical data to understand causes to better offer solutions. Research concerning sporting events and responsible consumption and production have long focused attention on environmental concerns (Collins et al., 2012). Considerations that relate to Goal 12 relate to adverse consumption habits that tie in with sporting events in a given locale. While many acknowledge consumption and production issues locally, and look to reduce habits to stop harm, the effects of travel are often overlooked as mobility to and from are less acknowl-edged, so this element of negative impacts appears overlooked (Cerezo-Esteve et al., 2022). Consumption from production concerning climate has been brought to the forefront of organizations such as FIFA, but they defer to the host countries and cities to adopt regulations to reduce their footprint (Fermeglia, 2017).

The United Nations Climate Change group published a call for action on climate with a specific focus on sports (United Nations Climate Change, 2017). Directly in line with Goal 13, the action framework puts forward a pledge to reduce emissions and promote responsible consumption (also in line with Goal 12). One concern here is the pledge and principles focus more on these issues locally, and while travel is acknowledged, it is not a major focus point of the report. Therefore, this study is important because it addresses the most crucial part of climate concern, that is, travel to and from as well as the mode of transport, which is the main con-tribution of carbon emissions as people seek to participate in and watch events in both domestic and international settings. While Hugaerts et al. (2021) acknowledge Goal 13 as a major challenge, empirical research needs to reinforce this conscious-ness with evidence that managers can use to guide decisions and inform partici-pants of their climate impact when deciding to travel to partake in or watch events.

This chapter aims to enhance awareness on the carbon footprint caused by par-ticipants' travelling to sports events and, based on their travel distance and style, suggests some very specific strategies on how to reduce this impact. Context and data for this chapter are based on the case of the Ultra-Trail World Tour (UTWT) event titled 100 Miles of Istria 2021 (100MoI-2021), held in Croatia. Through this chapter, we refer to strategies and recommendations that align with SDGs 12 and 13. For this, we discuss in this chapter how to find the optimal balance between the interest of event organizers who want their events to be international and the environmental concerns regarding CO_2 emissions related to travel. This chapter will now turn to the study context and empirical approach implemented to calculate event participants' travel CO_2 emissions. The results section will present the main findings and position several strategies that align with how event managers can aim to reduce total travel CO_2 emissions of participants. The conclusion summarizes this chapter's main findings and offers further recommendations and directions for future research so that contributions to research concerning travel to and from events can create new insights and impacts.

Study context

As an outdoor and off-road sport, trail running has seen growth in popularity over the past decade. This is due to the fact that athletes find it is possible to combine psycho-physical challenge and beautiful natural sceneries while participating in these sporting events (Botella-Carrubi et al., 2019; Myburgh & Kruger, 2021; Perić & Slavić, 2019). Other scholars have found that this sport helps contribute to the destination development of more rural locales as well (Lukoseviciute et al., 2022; Perrin-Malterre, 2018). Therefore, natural environment seems to be of crucial importance for trail runners. It is then assumed that participants would be concerned with and behave in environmentally friendly manners during the race (Newland et al., 2021; Myburgh & Kruger, 2021). However, it remains unknown how they travel and what CO_2 emissions their travel generate, and this is a crucial component that potentially compromises and thus contradicts the above points about an event where we assume that people would have a high level of environmental consciousness.

100MoI-2021 is the largest self-sufficiency trail running race in Istria, Croatia, and the Adriatic region (Figure 4.1). It is certified by the International Trail-Running Association (ITRA) and a part of the UTWT, a collection of the world's most prestigious ultra-trail races. The event is typically held in April each year, but due to COVID-19 in 2020 and 2021, the 2021 edition was held in September (Friday 10 to Sunday 12). The 2021 edition of the event attracted 988 competitors from 29 different countries. Competitors could compete in one of the five different courses (Red: 168 km and time limit 46 hours; Blue: 128 km and time limit 34 hours; Green: 67 km and time limit 15 hours; Yellow: 41 km and time limit 8 hours; and White: 20 km and time limit 4 hours). Social distancing measures put in place in 2021 meant that this edition of the event would not have any supplemental or accompanying programmes (e.g. kids' race and sports fair).

Methodology

Questionnaire and data collection

To collect data from those participating in the 100MoI-2021 event, a self-administered questionnaire was developed containing four parts altogether, of which only two were relevant for this study. The first relevant part was guided by the work of Triantafyllidis et al. (2018) and Wicker (2019) to shape 14 questions referring to

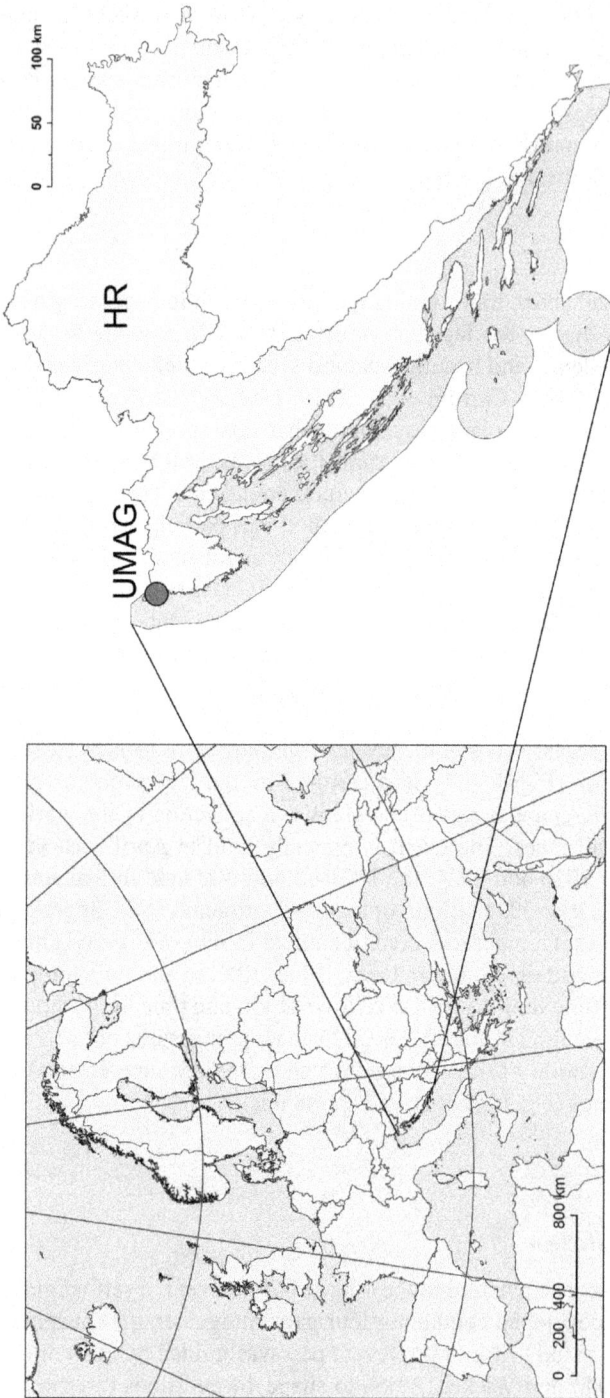

Figure 4.1 Location of the city of Umag in Croatia where the 100MoI-2021 race is held.

travel-related data aimed to capture and further calculate CO_2 emission values. The second relevant part of the questionnaire collected participants' socio-demographic data (age, gender, place of residence, length of living in the place of residence, education, and employment).

Trail runners were approached in person before the race started and asked if they are willing to complete the online questionnaire. A flier with a QR code leading to the electronic version of the questionnaire (available at www.1ka.si) was given to those who agreed to participate in the survey. Anonymity was guaranteed to all participants. Most of the participants responded to the questionnaire in a short period after the race. To increase the response rate, the event organizer sent an e-mail reminder for filling out the survey. In total, 754 responses were collected, of which 376 were acceptable for this study (38.1% of all trail runners who participated in the race).

The respondents' profiles are presented in Table 4.1. The sample is male dominated (65.9%), with 44.4% of international participants (mostly from Slovenia, Italy, Czech Republic, Germany, Hungary, Austria, and Poland). The average age of the respondents was 41.9 years, they are highly educated, and more than 50%

Table 4.1 Sample description

Variable		N	%
Gender	Male	244	64.89
	Female	132	35.11
Country of origin	Domestic – Croatia	209	55.59
	International	167	44.41
Education	Elementary	4	1.06
	Secondary	94	25.00
	Higher or more	278	74.94
Monthly (net) income	0–499	16	4.26
	500–999	49	13.03
	1,000–1,499	144	38.30
	1,500–1,999	49	13.03
	2,000–2,499	47	12.50
	2,500–2,999	20	5.32
	3,000–3,499	12	3.19
	3,500–3,999	9	2.39
	≥ 4,000	30	7.98
Transport	By car	341	90.69
	By airplane and rent-a-car	25	6.65
	On foot	10	2.66
Buffers	A: 0–100 km	85	22.61
	B: 101–300 km	156	41.49
	C: 301–700 km	82	21.81
	D: 701–1,500 km	36	9.57
	E: 1,501–3,100 km	12	3.19
	F: 3,101–6,300 km	2	0.53
	G: 6,301–12,700 km	3	0.80

of participants have a monthly net income between 1,000 and 2,000 EUR. The profiles (Table 4.1.) are in line with previous studies on running and other out-door or nature-based sports; these sports are also male dominated with highly edu-cated people (Getz & McConnell, 2014; Buning et al., 2016; Duglio & Beltramo, 2017; Perić et al., 2019; Myburgh & Kruger, 2021). More than 90% of participants arrived at the venue in a car, others arrived by combining airplane and car (6.6%) or on foot (2.7%). Almost three-quarters of the respondents (64.1%) had travelled less than 300 km to arrive at the venue, 21.8% had travelled between 301 and 700 km, 9.6% (36) had travelled between 701 and 1,500 km, and 3.2% (12) had travelled between 1,501 and 3,100 km, while only 1.3% (5) had travelled more than 3,101 km to arrive at the venue. In other words, more than 95% of participants arrived from places of residence less than 1,500 km away.

Data analysis

This chapter applied the carbon footprint (CF) method to convert trail runners' travel-related data into specific CO_2 emission values expressed in kilograms (kg CO_2). Travel distance, type of vehicle, its engine and consumption, and travel party (travelling alone or in company) were considered when calculating the CO_2 emis-sion (Scrucca et al., 2021; eea.europa.eu, 2017; Grofelnik, 2010). Travel distance (in km) assumed return travel, and, therefore, it was doubled. Next, using ArcGIS Map tools for analysing and mapping the CF of participants, the gravity zone of the event was divided into geographical "buffer zones" of different CF intensi-ties. Finally, depending on the characteristics of participants belonging to specific "buffer zones" (e.g. means of transport and travel style), some practical recommen-dations to event organizers and policymakers are proposed.

Results and interpretations

Table 4.2 and Figure 4.2 show that transport sharing has a significant impact on the reduction of CF per competitor. This is particularly the case with the first four buffers A, B, C, and D (0–100 km, 101–300 km, 301–700 km, and 701–1,500 km, respectively), where, on average, every other trail runner shares transportation (46–56%). Primarily, people decide to share transport because of increased costs of travelling by car (petrol and tools) (Standing et al., 2019). As a result, the average real CF per trail runner in the first four buffers is 21–29% lower than theoretical estimates (i.e. without transport sharing). In the next three long-distance buffers E, F, and G (1,501–3,100 km, 3,101–6,300 km, and 6,301–12,700 km, respectively), trail runners used airplane to arrive at the destination, and ride sharing is related only to travelling from the airport to the venue. Therefore, in these three buffers, the average real CF per trail runner decreased very slightly (approximately 1%). On the aggregate level (i.e. for the whole sample), the average real CF per trail run-ner is 18% lower than theoretical estimates.

While approximately 47.3% of trail runners shared transportation, it is interest-ing to notice that the relationship between the number of runners using vehicle

Table 4.2 Travel CF per buffer zone

Buffer zone	N	Theoretical ∑ CF per buffer*		Ride sharing			Corrected ∑ CF per buffer**	
		∑ CF (kg CO₂)	Average CF per competitor (kg CO₂)	Number of competitors	Number of rides	Avg. number of competitors per vehicle	Real CF on 100 MOI (kg CO₂)	Avg. real CF per competitor (kg CO₂)
A	85	876.36	10.31	39	86	2.21	687.95	8.09
B	156	5,966.96	38.25	75	188	2.51	4,256.20	27.28
C	82	6,347.25	77.41	39	101	2.59	4,640.69	56.59
D	36	5,511.97	153.11	20	54	2.70	3,901.03	108.36
E	12	4,045.20	337.10	3	6	2.00	4,003.43	333.62
F	2	1,275.45	637.72	1	2	2.00	1,268.40	634.20
G	3	4,810.59	1,603.53	1	2	2.00	4,789.37	1,596.46
∑	376	28,833.78	76.69	178	439	2.47	23,547.08	62.63

Note: A = 0–100 km, B = 101–300 km, C = 301–700 km, D = 701–1,500 km, E = 1,501–3,100 km, F = 3,101–6,300 km, and G = 6,301–12,700 km.

* Calculated without taking into account transport sharing.
** Calculated taking into account transport sharing.

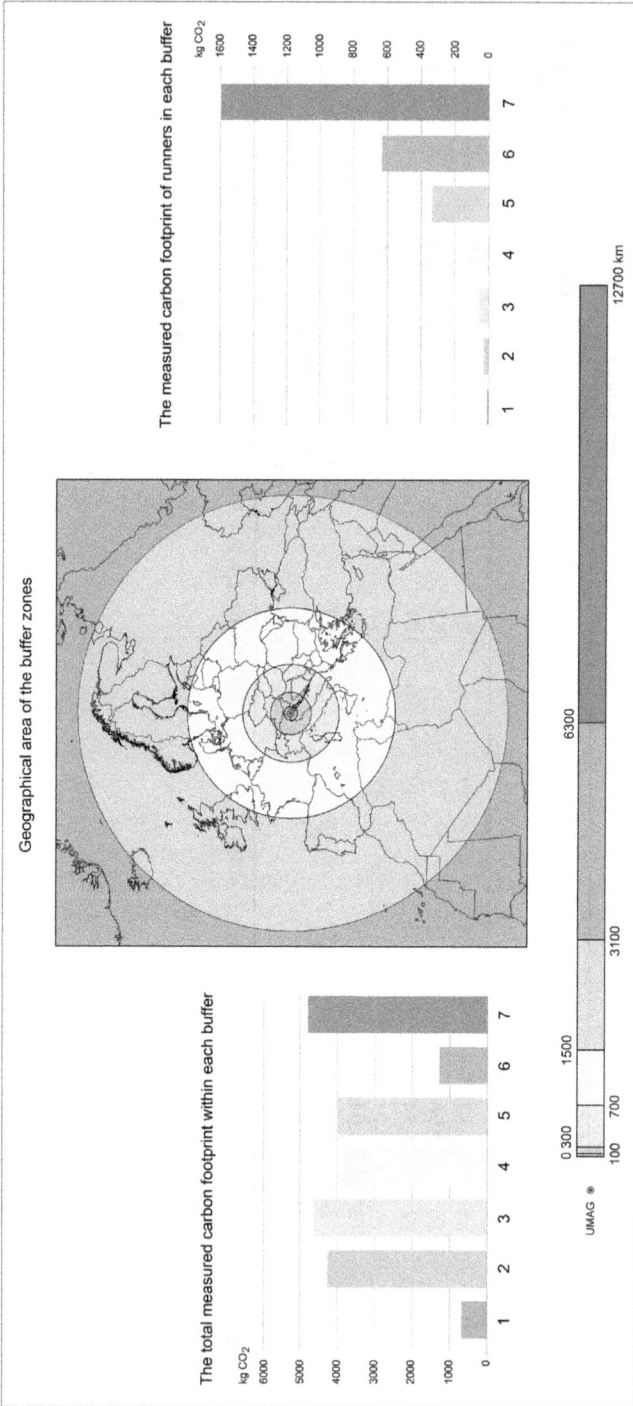

Figure 4.2 Geographic distribution of CF between buffers and personal CF of runners in each buffer.

does not show regularity with the increase in distance travelled (Table 4.3). Furthermore, it seems that trail runners with a lower level of income tend to share transportation more regularly. The same is with the level of education (the lower the education level, the higher the level of ride sharing), but this is not a surprise because educational level is very often correlated with income (Perić et al., 2019). On the other hand, the age and gender of the trail runners do not show any regularity in relation to ride sharing.

Since the aim of this chapter is to provide very specific recommendations to event organizers regarding travel CF reduction, the next step was oriented to find alternative scenarios of collective transport (i.e. bus, minibus, or van) and recalculate the average CF per trail runner. First, two buffers with two cities each were selected based on the criteria that only cities of departure that had a sufficient number of competitors at the 100MOI-2021 can be taken into account for the opportunity to provide collective transport. In the first buffer A (0–100 km), there are Pula and Rijeka (both in Croatia), and in the second buffer B (101–300 km) are Ljubljana (in Slovenia) and Zagreb/Samobor (in Croatia, joined due to proximity). Because it is not realistic that all trail runners will be willing to participate in an alternative scenario of collective transport, several scenarios are provided (100%, 90%, 75%, or 50% of trail runners will join the initiative of collective transport, and bus/minibus/van options are provided depending on the number of trail runners from the particular city). While the significant savings in travel CF can be achieved by using organized collective transport (Tables 5–7), the key point is to find the minimum number of trail runners who, if they choose collective transport, will have lower levels of travel CF compared to the individual travel CF of tail runners who share a car with another competitor (this is used as a benchmark because, on average, at the 100MOI-2021, two trail runners share a car). Hence, the last column shows the hypothetical scenario with the smallest turnout of the number of runners that is environmentally acceptable for the organization of collective transport of runners. For instance, from the environmental

Table 4.3 Ride sharing sample description

Ride sharing – number of competitors in the vehicle	Number	Avg. ride distance (km)*	Income (avg. category)	Age (years)	Education (avg. category)	Gender, M (N)	Gender, M (%)
1 (Not sharing the ride)	198	472.832	4.09	42.0	2.737	137	69.2
2	123	434.984	4.29	41.7	2.756	81	65.9
3	30	298.367	3.40	42.7	2.667	12	40.0
4	22	444.136	3.36	40.6	2.636	12	54.5
5	3	275.000	3.33	44.8	2.333	2	66.7
∑ (Ride sharing competitors; 1 + 2 + 3 + 4)	178	410.393	4.01	41.8	2.719	107	60.1

* The sum in column three cannot be added directly because some racers overlap in sharing transportation.

perspective, it is worth organizing a collective transport for six trail runners by a minibus or four trail runners by a van from Pula (see Table 4.4). In the case of Zagreb/Samobor, it is worth organizing a collective transport for 10 trail runners by a bus (see Table 4.6). In all these cases, the average travel CF per trail runner is lower than it would be if travelling by a shared car (two trail runners in each car). These numbers (from the last column) are very small, implying that the organization of collective transport from the ecological point of view is worthwhile.

In Tables 4–6 in the penultimate column, the turnout of 50% of runners is shown (in relation to the number of completed surveys from the starting point at

Table 4.4 Alternative scenarios of travel CF per trail runner (kg CO_2) depending on the departure places, means of transport, and occupancy of the vehicle, departure from Pula and Rijeka

Item	Personal car		Minibus (capacity 22 runners) or a van (capacity 8 runners)					
	1 runner	2 runners	Minibus turnout 100% (N 16)	Minibus turnout 90% (N 14)	Minibus turnout 75% (N 12)	Minibus turnout 50% (N 8)	Van turnout 50% (N 8)	Minimum positive turnout (compared to a car with two runners)**
CF per runner (kg CO_2), Pula (N 16)	13.924	6.962*	2.230	2.549	2.974	4.460	3.457	6 runners – minibus (CF 5.947), 4 runners – van (CF 6.914)
CF per runner (kg CO_2), Rijeka (N 13)	16.079	8.039*	3.170	3.434	4.121	5.887	4.562	6 runners – minibus (CF 6.868), 4 runners – van (CF 7.984)

Table 4.5 Alternative scenarios of travel CF per trail runner (kg CO_2) depending on the departure places, means of transport, and occupancy of the vehicle, departure from Ljubljana

Item	Personal car		Minibus (capacity 22 runners)				
	1 runner	2 runners	Turnout 100% (N 17)	Turnout 90% (N 15)	Turnout 75% (N 13)	Turnout 50% (N 9)	Minimum positive turnout (compared to a car with two runners)**
CF per runner (kg CO_2), Ljubljana (N 17)	21.715	10.857*	3.273	3.710	4.281	6.183	6 runners (CF 9.275)

Table 4.6 Alternative scenarios of travel CF per trail runner (kg CO_2) depending on the departure places, means of transport, and occupancy of the vehicle, departure from Zagreb

Item	Personal car		Bus (capacity 50 runners)				
	1 runner	2 runners	Turnout 100% (N 48 + 48, two buses)	Turnout 90% (N 43 + 43, two buses)	Turnout 75% (N 36 + 36, two buses)	Turnout 50% (N 43, one bus)	Minimum positive turnout (compared to a car with two runners)**
CF per runner (kg CO_2), Zagreb (N 82) + Samobor (N 14)	46.744	23.372*	4.679	5.224	6.239	5.224	10 runners (CF 22.461)

* Number to be compared with the last column that shows the minimum positive turnout.
** The last column shows the minimum numbers of runners where the CF of organized transport is lower than the CF of a car with two runners.

Table 4.7 Reduction scenario per buffer at a response rate of 50% of organized transport compared to the CF produced by sharing a car

Buffer	Measured absolute reduction of CF per runner due to sharing of personal transport (kg CO_2)	Measured relative reduction of CF per runner due to sharing of personal transport (%)	Absolute values of possible additional CF reduction by implementing organized transport* (kg CO_2)	Relative values of possible additional CF reduction by implementing organized transport* (%)
A	−2.216	−21.49	−59.594	−9.24
B	−10.967	−28.67	−280.264	−19.04
C	−20.812	−26.89		
D	−44.748	−29.23		

* Travel CF reduction scenario with buffers A and B at a response rate of 50% of runners taking organized transport compared to travel CF produced by a car shared by two persons (the number of runners surveyed at 100MOI-2021).

100MOI-2021). It is visible that with that scenario that takes the lowest turnout and compares it with the CF of runners who hypothetically share transportation in a car with another runner, the CF of organized transport is between 43% and 78% lower. Also, it is important to mention that this is the worst scenario, and scenarios with 75%–100% of turnout have an even greater reduction of CF per passenger.

Besides the relative reduction of travel CF per trail runner due to sharing of personal transport, relative values of possible additional travel CF reduction by implementing organized collective transport (using buses, minibuses, and vans) are calculated for buffers A and B (Table 4.7). Other buffers have not been taken into account because there were not cities (i.e. departure places) with enough trial runners to do the organized transport. While the average real travel CF per trail runner in buffers A–D would be reduced by 21–29% if they share transport (i.e. two persons in a car), travel CF savings will be even higher if collective transport would be organized.

In case 50% of turnout of trail runners from the first (cities of Pula and Rijeka) and second buffers (cities of Ljubljana and Zagreb) join the initiative and arrive at the event by organized transport (bus, minibus, or van), travel CF will be additionally reduced by 9% and 19% respectively. In this case, a total travel CF reduction at the level of the whole 100MOI-2021 event is projected as 1,777 kg CO_2 or 7.55% when compared to the real travel CF footprint per participant where almost half of the participants shared transport as a couple in a private car.

Conclusions and recommendations

If properly developed and managed, sports tourism can play an important role in achieving various SDGs. However, when we consider consumption, it must not only refer to consumption at the venue or in the host community but also refer to the mode of travel as a consumption practice from the place of origin. Travel to and from the event is where the climate impact is most important. As we live in an increasingly globalized world and with people wanting to travel to different places and environments to participate in events, this makes it difficult for managers to reduce their carbon footprint. Ultimately, the carbon footprint goes back to individual participants as they decided to travel a long or short distance to participate. The challenge is how to reconcile the personal interests of the competitors, the interests of the organizers, and the general social interest in reducing the impact on the environment.

This research showed that if a certain amount of effort is invested in the process of event preparation and management, it is possible to make a step forward in reducing the event's CF and making it more environmentally friendly. The results of the presented CF reduction models in this research are not enough to solve the global CF problem, which is ultimately the question of technological nature, but indicate that thoughtful event management can contribute to reducing the impact on the environment and getting closer to meeting the global goals of sustainable development. Elaborated models indicate that larger regional sports and recreation events that gather a larger number of participants should be developed within the first two buffer zones (up to 300 km distance). At such events of a sports-recreational nature, the strategies of organizing shared transportation and sharing personal transportation of participants could be applied. In other words, besides ride sharing, which is a beneficial strategy for travel CF reduction, event organizers could organize collective transport (such as buses, minibuses, and vans) for participants residing in cities within the first two buffer zones. Such strategies, in addition to the tourist recreational aspect, could be used as qualifications for top-ranked events. Therefore, higher levels of competition could be of a sports-competitive nature, with a smaller number of participants coming from greater distances. A reduction in the number of participants at such top-ranked events could provide more time to organizers to rework and rearrange competitors' schedules and to propose alternative, more environmentally friendly ways of arriving at the event to the participants. Additionally, if the longer distance buffers do not have a sufficient number of competitors (located in cities of departure) for the organization of buses/minibuses/vans, it is desirable to encourage the sharing of contacts among registered runners. Event organizers could design an online platform for self-organization and sharing of transport among runners in order to increase the number of runners who share transport. In order to reduce the traffic footprint, and this is eligible for all buffers, event organizers could motivate participants to use shared/collective transport by offering promotional discounts or benefits during their stay at the event itself.

To summarize, this chapter provided a novel perspective on how we understand the environmental impacts of active sports tourists' event-related travel CF. It also offered new solutions for developing appropriate strategies and incentives to encourage attendees' pro-environmental behaviours. These efforts might support better event management from the environmental perspective and contribute to key SDGs.

References

Ahmed, T. S. A. A. (2017). A triple bottom line analysis of the impacts of the Hail International Rally in Saudi Arabia. *Managing Sport and Leisure*, 22(4), 276–309. https://doi.org/10.1080/23750472.2018.1465841

Boluk, K. A., Cavaliere, C. T., & Higgins-Desbiolles, F. (2019). A critical framework for interrogating the United Nations Sustainable Development Goals 2030 Agenda in tourism. *Journal of Sustainable Tourism*, 27(7), 847–864. https://doi.org/10.1080/09669582.2019.1619748

Botella-Carrubi, D., Currás Móstoles, R., & Escrivá-Beltrán, M. (2019). Penyagolosa trails: From ancestral roads to sustainable ultra-trail race, between spirituality, nature, and sports. A case of study. *Sustainability*, 11(23), 6605. https://doi.org/10.3390/su11236605

Cerezo-Esteve, S., Inglés, E., Segui-Urbaneja, J., & Solanellas, F. (2022). The environmental impact of major sport events (Giga, Mega and Major): A systematic review from 2000 to 2021. *Sustainability*, 14(20), 13581. https://doi.org/10.3390/su142013581

Collins, A., Munday, M., & Roberts, A. (2012). Environmental consequences of tourism consumption at major events: An analysis of the UK stages of the 2007 tour de France. *Journal of Travel Research*, 51(5), 577–590. https://doi.org/10.1177/0047287511434113

Cooper, J. A., & McCullough, B. P. (2021). Bracketing sustainability: Carbon footprinting March Madness to rethink sustainable tourism approaches and measurements. *Journal of Cleaner Production*, 318, 128475. https://doi.org/10.1016/j.jclepro.2021.128475

Deery, M., Jago, L., & Fredline, L. (2004). Sport tourism or event tourism: Are they one and the same? *Journal of Sport & Tourism*, 9(3), 235–245. https://doi.org/10.1080/1477508042000320250

Dolf, M., & Teehan, P. (2015). Reducing the carbon footprint of spectator and team travel at the University of British Columbia's varsity sports events. *Sport Management Review*, 18(2), 244–255. https://doi.org/10.1016/j.smr.2014.06.003

Energy Efficiency and Specific CO_2 Emissions. (2017). Retrieved on 22nd December, 2022, from www.eea.europa.eu/data-and-maps/indicators/energy-efficiency-and-specific-co2-emissions/energy-efficiency-and-specific-co2-9

Fermeglia, M. (2017). The show Must Be Green: Hosting mega-sporting events in the climate change context. *Carbon & Climate Law Review*, 11(2), 100–109. www.jstor.org/stable/26353858

Fodness, D. (2017). The problematic nature of sustainable tourism: Some implications for planners and managers. *Current Issues in Tourism*, 20(16), 1671–1683. https://doi.org/10.1080/13683500.2016.1209162

Grofelnik, H. (2010). Ecological footprint of road traffic on Cres-Lošinj Archipelago. *Geoadria*, 15, 269–286, https://doi.org/10.15291/geoadria.191

Grofelnik, H., Perić, M., & Wise, N. (2020). Applying carbon footprint method possibilities to the sustainable development of sports tourism. *WIT Transactions on Ecology and the Environment*, 248, 153–163, https://doi.org/10.2495/ST200131

Haase, E. (2022). Driving the environmental extra mile – car sharing and voluntary carbon dioxide offsetting. *Transportation Research Part D: Transport and Environment*, 109, 103361. https://doi.org/10.1016/j.trd.2022.103361

Hall, C. M. (2019). Constructing sustainable tourism development: The 2030 Agenda and the managerial ecology of sustainable tourism. *Journal of Sustainable Tourism*, 27(7), 1044–1060. https://doi.org/10.1080/09669582.2018.1560456

Hugaerts, I., Scheerder, J., Helsen, K., Corthouts, J., Thibaut, E., & Könecke, T. (2021). Sustainability in participatory sports events: The development of a research instrument and empirical insights. *Sustainability*, 13(11), 6034. https://doi.org/10.3390/su13116034

Lenzen, M., Sun, Y. Y., Faturay, F., Ting, Y. P., Geschke, A., & Malik, A. (2018). The carbon footprint of global tourism. *Nature Climate Change*, 8(6), 522–528. https://doi.org/10.1038/s41558-018-0141-x

Liu, D., Hautbois, C., & Desbordes, M. (2017). The expected social impact of the Winter Olympic Games and the attitudes of non-host residents toward bidding: The Beijing 2022 bid case study. *International Journal of Sports Marketing and Sponsorship*, 18(4), 330–346. https://doi.org/10.1108/IJSMS-11-2017-099

Lukoseviciute, G., Pereira, L. N., & Panagopoulos, T. (2022). Assessing the income multiplier of trail-related tourism in a coastal area of Portugal. *International Journal of Tourism Research*, 24(1), 107–121. https://doi.org/10.1002/jtr.2487

Moyle, B. D., Weaver, D. B., Gössling, S., McLennan, C. L., & Hadinejad, A. (2022). Are water-centric themes in sustainable tourism research congruent with the UN sustainable development goals? *Journal of Sustainable Tourism*, 30(8), 1821–1836. https://doi.org/10.1080/09669582.2021.1993233

Myburgh, E., & Kruger, M. (2021). The sky is the limit: A motivation and event attribute typology of trail runners. *Managing Sport and Leisure*, 1–20. https://doi.org/10.1080/23750472.2021.1987302

Newland, B. L., Aicher, T. J., Davies, M., & Hungenberg, E. (2021). Sport event ecotourism: Sustainability of trail racing events in US National Parks. *Journal of Sport & Tourism*, 25(2), 155–181. https://doi.org/10.1080/14775085.2021.1902374

Nunkoo, R., Sharma, A., Rana, N. P., Dwivedi, Y. K., & Sunnassee, V. A. (2021). Advancing sustainable development goals through interdisciplinarity in sustainable tourism research. *Journal of Sustainable Tourism*, 1–25. https://doi.org/10.1080/09669582.2021.2004416

Pedauga, L. E., Pardo-Fanjul, A., Redondo, J. C., & Izquierdo, J. M. (2022). Assessing the economic contribution of sports tourism events: A regional social accounting matrix analysis approach. *Tourism Economics*, 28(3), 599–620. https://doi.org/10.1177/1354816620975656

Pereira, R. P. T., Filimonau, V., & Ribeiro, G. M. (2019). Score a goal for climate: Assessing the carbon footprint of travel patterns of the English Premier League clubs. *Journal of Cleaner Production*, 227, 167–177. https://doi.org/10.1016/j.jclepro.2019.04.138

Perić, M. (2018). Estimating the perceived socio-economic impacts of hosting large-scale sport tourism events. *Social Sciences*, 7(10), 176. https://doi.org/10.3390/socsci7100176

Perić, M., & Slavić, N. (2019). Event sport tourism business models: The case of trail running. *Sport, Business and Management: An International Journal*, 9(2), 164–184. https://doi.org/10.1108/SBM-05-2018-0039

Perić, M., Wise, N., Heydari, R., Keshtidar, M., & Mekinc, J. (2021). Getting back to the event: COVID-19, attendance and perceived importance of protective measures. *Kinesiology*, 53(1), 12–19. https://doi.org/10.26582/k.53.1.2

Perrin-Malterre, C. (2018). Tourism diversification process around trail running in the Pays of Allevard (Isère). *Journal of Sport & Tourism*, 22(1), 67–82. https://doi.org/10.1080/1 4775085.2018.1432410

Polcsik, B., & Perényi, S. (2022). Residents' perceptions of sporting events: A review of the literature. *Sport in Society*, 25(4), 748–767. https://doi.org/10.1080/17430437.2021.1982899

Rasoolimanesh, S. M., Ramakrishna, S., Hall, C. M., Esfandiar, K., & Seyfi, S. (2020). A systematic scoping review of sustainable tourism indicators in relation to the sustainable development goals. *Journal of Sustainable Tourism*, 1–21. https://doi.org/10.1080/0966 9582.2020.1775621

Ritchie, B. W., Chien, P. M., & Shipway, R. (2020). A leg(acy) to stand on? A non-host resident perspective of the London 2012 Olympic legacies. *Tourism Management*, 77, 104031. https://doi.org/10.1016/j.tourman.2019.104031.

Scrucca, F., Barberio, G., Fantin, V., Porta, P. L., Barbanera, M. (2021). Carbon footprint: concept, methodology and calculation. In S. S. Muthu (Ed.), *Carbon footprint case studies environmental footprints and eco-design of products and processes* (pp. 1–31). Singapore: Springer, Online ISBN 978-981-15-9577-6. https://doi.org/10.1007/978-981-15-9577-6_1

Sinha, R., Olsson, L. E., & Frostell, B. (2019). Sustainable personal transport modes in a life cycle perspective – public or private? *Sustainability*, 11(24), 7092. https://doi. org/10.3390/su11247092

Standing, C., Standing, S., & Biermann, S. (2019). The implications of the sharing economy for transport. *Transport Reviews*, 39(2), 226–242. https://doi.org/10.1080/01441647.201 8.1450307

Thormann, T. F., Wicker, P., & Braksiek, M. (2022). Stadium travel and subjective well-being of football spectators. *Sustainability*, 14(12), 7278. https://doi.org/10.3390/ su14127278

Tomino, C. A., & Perić, M. (2022). Sport-tourism running events in the post-Covid-19 World: Any sign of change? *Academica Turistica – Tourism and Innovation Journal*, 15(1), 135–147. https://doi.org/10.26493/2335-4194.15.135-147

Triantafyllidis, S., Ries, R. J., & Kaplanidou, K. (2018). Carbon dioxide emissions of spectators' transportation in collegiate sporting events: Comparing on-campus and off-campus stadium locations. *Sustainability*, 10(1), 241. https://doi.org/10.3390/su10010241

United Nations Framework Convention on Climate Change (UNFCCC). (2017). *Sports for climate action framework*. New York, USA: United Nations.

Van Fan, Y., Perry, S., Klemeš, J. J., & Lee, C. T. (2018). A review on air emissions assessment: Transportation. *Journal of Cleaner Production*, 194, 673–684. https://doi. org/10.1016/j.jclepro.2018.05.151

Wallstam, M., & Kronenberg, K. (2022). The role of major sports events in regional communities: A spatial approach to the analysis of social impacts. *Event Management*, 26(5), 1025–1039. https://doi.org/10.3727/152599522X16419948390781

Weed, M. (2020). The role of the interface of sport and tourism in the response to the COVID-19 pandemic. *Journal of Sport & Tourism*, 24(2), 79–92. https://doi.org/10.1080 /14775085.2020.1794351

Weed, M., & Bull, C. (2009). *Sports tourism: Participants, policy and providers* (2nd ed.). Oxford: Elsevier.

Wicker, P. (2019). The carbon footprint of active sport participants. *Sport Management Review*, 22(4), 513–526. https://doi.org/10.1016/j.smr.2018.07.001

World Tourism Organization. (2019). *Sport tourism and the sustainable development goals (SDGs)*. Madrid: UNWTO. https://doi.org/10.18111/9789284419661.

Zhang, C., Zhou, X., Zhou, B., & Zhao, Z. (2022). Impacts of a mega sporting event on local carbon emissions: A case of the 2014 Nanjing Youth Olympics. *China Economic Review*, 73, 101782. https://doi.org/10.1016/j.chieco.2022.101782

Zhou, R., Kaplanidou, K., & Wegner, C. (2021). Social capital from sport event participation: Scale development and validation. *Leisure Studies*, 40(5), 612–627. https://doi.org/ 10.1080/02614367.2021.1916832

5 Developing transnational cycling destinations with the help of EU funds: A truly sustainable approach?

Hristo Andreev and Miha Bratec

Introduction – defining cycle tourism

Cycle tourism is a specialized form of tourism that combines the physical activity of cycling with travel and leisure. There are various definitions of cycle tourism in the literature because of the diversity and ambiguity of the cycle tourism concept (Lamont, 2009). Cycle tourism varies by plenty of factors, such as motives, types of cycling, location, duration, type of bicycle, group size, fitness level, goals, budget, support, weather, road conditions, and other factors (Andeev & Zopiatis, 2022). Dickinson and Lumsdon (2010) defined cycle tourists as those whose main focus of travel is cycling and who stay overnight in a destination (using lodging facilities like hotels, hostels, camping areas, stealth camping, etc.). The authors argue that non-overnight stayers are cycle excursionists or recreational cyclists, not tourists. Similar views are expressed by Simonsen and Jorgenson (1998).

Active cycle tourism, according to Faulks et al. (2008), is "recreational trips, either overnight or for a day, away from home, where leisure cycling is a central and important part of the trip." This definition includes both single-day and multi-day bike trips as "cycle tourism." Pavione and Pezzetti (2016) add that "cycle tourists" are tourists of any nationality who use bicycles to get around and make cycling a big part of their vacation. Most researchers agree that a cycle tourist must be away from home for 24 hours. However, non-overnight stayers outnumber overnight stayers and equally boost local economies. As a result, they should be included in the cycle tourist category (Weed et al., 2014).

Lamont (2009) gives a technical definition of cycle tourism as "trips involving a minimum distance of 40 kilometers from a person's home and an overnight stay (for overnight trips) or a minimum non-cycling round trip component of 50 kilometers and a minimum four-hour period away from home (for day trips), of which cycling is the main purpose for a holiday, recreation, leisure, or competition."

Nature and trends of cycle tourism

Most cycle tourists travel in small groups. For many people, cycling's social aspect motivates participation in cycle tourism (Pavione & Pezzetti, 2016). Team cycling gives participants a sense of belonging. Because cycling events shape their

DOI: 10.4324/9781003384786-5

social identities, cyclists often socialize during cycling and cycling-related events (Coghlan, 2012). Cycle tourists also value socializing with other cyclists (Faulks et al., 2008).

Cycle tourism participation depends on gender, location, and age (Gibson & Chang, 2012). Cycle tourism is a male-dominated activity (Rejón-Guardia et al., 2018). Males outnumber females by a big margin in most countries (Rejón-Guardia et al., 2018; Buning et al., 2019). Female cyclists emphasize mostly the educational, social, and safety aspects of cycle tourism, while male cyclists are driven mainly by the physical aspects (Gibson & Chang, 2012). Japanese, German, and Dutch people cycle the most due to cycling heritage, terrain, infrastructure, and cyclist safety. Cycling levels in a particular region cannot determine the number of cyclists participating in cycling tours. Probabilistically, however, the highest percentage of cyclists in the general population will result in the largest number of bike tourists in the region (Goel et al., 2021).

Road hazards are the number one barrier preventing people from cycling. Cyclists avoid busy and high-speed roads for safety reasons. More experienced cyclists can navigate busier roads. Intermediate cyclists prefer bike lanes, paths, and roads with moderate traffic (Goel et al., 2021). Cyclists prefer quiet, low-traffic routes. Most road cyclists prefer smooth, clean, low-traffic roads (Lee & Huang, 2014). Cyclists and bike tourists tend to avoid heavy motorized traffic. Most nations struggle to build traffic-free bike paths in cities and rural areas. To attract cycle tourists, rural routes with low traffic must be prioritized (Downward & Lumsdon, 2001). Recreational cyclists and cycle tourists prefer group rides not only for socialization but also for safety reasons. Traffic safety is crucial for the positioning and branding of countries that aspire to target cycle tourists (Pucher & Buehler, 2016).

Due to limited luggage capacity, cycle tourists are more likely to buy products from rural and remote areas. Weight slows a bike, especially in hilly, mountainous areas (Pavione & Pezzetti, 2016). Cycling lets tourists see the whole region leisurely. Cycle tourists spend a lot of time traveling between stops. Due to the physical strain of cycling, they stop for rest, food, and hydration. Cycle tourists do not have time, space, or equipment to cook (Simonsen & Jorgenson, 1998). Due to longer trips and higher physical demands, cycle tourists who travel long hours and large distances tend to spend more on food and drinks. Therefore, their daily expenditure increases significantly. Moreover, less experienced cyclists anticipate stopping more frequently for food and refreshments (Weed et al., 2014).

The aforementioned infrastructure preferences are related mostly to road-based cycle tourism, which is also the most dominant in the market. Other types of cycle tourism use roads to a lesser extent and are less concerned with traffic. Mountain bikers prefer off-road trails (Buning et al., 2019). Gravel biking is another growing cycling trend. Gravel biking and gravel cycle tourism, which involve exploring routes off-road, have received very little interest from the scientific community so far (Legan, 2017). Finally, most cycle tourists prefer paved, low-traffic roads. Rural gravel routes and mountainous trails offer many gravel and mountain bike tourism opportunities.

Cycling destination conceptualization and characteristics

Location, accessibility, geomorphology, weather, infrastructure, cultural, and natural resources naturally affect cycle tourism. Tourists visit cycle tourism destinations for the cycling and the destination's landscapes, agriculture, cultural heritage, traditions, and culture (Pavione & Pezzetti, 2016). Numerous factors affect cycle tourism destination attractiveness. Most factors focus on tarmac-based cycle tourism, but not all of them apply to all cyclists. The main factors that determine cycle tourism destinations are the following:

- Bicycle facilities (safe roads, paths, signs, and bike parking)
- Extensive networks of gravel roads and off-road trails for gravel and mountain bike tourists
- Quality and variety of bicycle-friendly accommodations, restaurants, and services
- Cycling attractions, such as thematic routes, cycling landmarks, and historical sites
- High-quality information for cycle tourism, like promotional videos and routes mapped in digital geoinformation platforms
- Environmental attractiveness and scenery
- Appropriate weather conditions for cycling
- Organized bicycle tours and guides
- Cycling heritage based on local cycling culture
- Bicycle races and events, such as the organization and hosting of grand tours, gran fondos, and sportives.

Separate cycling facilities along busy roads and intersections and traffic calming in most residential areas can increase cycling. Cyclists in the Netherlands, Denmark, and Germany have wide road rights, bike parking, public transportation integration, and thorough traffic education and training. These nations also hold many bike promotion events. A combination of such efforts creates a cycling culture (Pucher & Buehler, 2008).

Cycle tourists require several amenities. Equipment, spare parts, bicycle servicing, and secure bicycle storage are crucial. Long-term cycle tourists usually use expensive bicycles and pay more for maintenance and storage (Rejon-Guardia et al., 2018). The literature emphasizes bicycle-friendly accommodations (Bakogiannis et al., 2020). However, it varies from segment to segment. Some groups prefer hostels and camping (Buning & Gibson, 2016). Cyclists, known as "nomadic cycle tourists," take long trips and change accommodations frequently (Weed et al., 2014). Local and out-of-town day travelers prefer vacation lodging with bicycle-friendly features and services like rentals, transportation, maintenance, and support (Bakogiannis et al., 2020). In conclusion, cycle tourism destinations must offer a variety of accommodations and services to meet the needs of cycle tourists.

Attractions add value to cycle tourism beyond cycling. Natural attractions include forests, lakes, rivers, beaches, mountains, and waterfalls. Man-made attractions include towns, historical sites, monuments, and shops (Lee & Huang, 2014).

Weather greatly affects cycle tourists. Some places are good for cycle tourism depending on the season and weather (Simonsen & Jorgenson, 1998). Most countries' low seasons – fall and spring – are ideal for cycle tourism (Pavione & Pezzetti, 2016). Thus, cycle tourism appeals to destination managers trying to reduce seasonality and build year-round tourist destinations.

Cycle tourists are happier when stores, restaurants, cafes, hotels, bike repair shops, bicycle rentals, tour guides, and public transportation are available and are cycling-friendly. Cyclists need shelter, food, water, and safety. Staying overnight requires bicycle-friendly hotels, hostels, vacation rentals, bed and breakfasts, and campsites (Bakogiannis et al., 2020). It is also crucial to have a safe space for bike parking and storage. Not all cyclists bring their own bikes to a destination. Thus, cycle tourism destinations need bike rental services. For shorter distances and fewer days in destinations that are not accessible by bicycle (i.e., islandic areas or faraway destinations), some people prefer to rent bikes nearby instead of packing and transporting their bikes on aeroplanes, trains, and buses (Lee & Huang, 2014).

Destination satisfaction is also influenced by tour operators and tour organizers who can offer guided tours, transportation for equipment and luggage, and support services (Rejon-Guardia et al., 2018). Last but not least, cycle tourists prefer public transportation services that allow for easy bicycle transportation, such as trains, buses, and aeroplanes (Lee & Huang, 2014).

Routes and their characteristics play a crucial role at cycling destinations. When a destination offers a variety of route lengths and terrains, such as circular or out-and-back day routes on quiet roads and overnight trails with a variety of lodging options along the routes, it becomes more alluring (Bakogiannis et al., 2020; Buning et al., 2019).

Cycling tradition, history, and landmarks are very important for cycle tourism destinations. Italy, France, and Spain, which host pro cycling events like Tour de France, Giro d'Italia, and La Vuelta, use their cycling cultural heritage to attract tens of thousands of active and passive cycle tourists each year (Weed et al., 2014; Varnajot, 2020). In Flanders, Belgium, cycle tourists can ride the exact route of the Tour of Flanders that professional cyclists will ride the day before the event (Derom & Ramshaw, 2016).

Cycle tourists need digital and/or printed maps and visuals with information on cycling routes, lodging, drinkable water fountains, restrooms, shops, and bike storage, as well as the terrain, grading, and surface of routes and trails (Buning et al., 2019). Countries have promoted their bike tours through various media and information sources. Before the Internet, cycle tourism destinations were advertised mainly with brochures, printed cycling maps, event calendars, travel guides, and books (Simonsen & Jorgenson, 1998). Smartphones and GPS devices like bike computers are making interactive maps and Geographic Information Systems (GIS) the most popular geoinformation sources for cycle tourists. Interactive digital cycling maps show distance, elevation profile, and point-to-point navigation. There is plenty of information about cycle tourism online. Destinations and local tourism organizations produce and distribute promotional material. However, most promotional content is user-generated on social media, blogs, forums, and online videos. Thus, cycle tourism influencers play a major role in promoting cycling destinations (Lőrincz et al., 2020).

Best practices for sustainable cycle tourism development and its benefits

Cycle tourism encourages local economic and entrepreneurial growth. Food, lodging, bicycle rentals, information, and wellness services like yoga and spas also benefit (Bakogiannis et al., 2020). Cycle tourism is making tourist destinations more competitive by boosting the destination's offer portfolio (Jurišić et al., 2018). Cycle tourism helps rural areas develop economically and makes them more well-known (Pavione & Pezzetti, 2016). Cycle tourism may be combined with wine tourism (Walsh, 2019), gastronomy, wellness, cultural, and other specialized forms of tourism (Jurišić et al., 2018). Renowned cycling competitions like the Giro d'Italia, Tour de France, Tour Down Under, and Tour of Flanders can encourage both active and passive cycle tourism (Shipway et al., 2016). Destination countries can also host gran fondos, sportives, and charity events related to cycling. These events can benefit local economies and society.

One of cycle tourism's main strengths is that it has the potential to significantly advance rural development. Old rural roads and abandoned railroads that can be converted into bike paths are examples of underused infrastructure that may be revived and utilized for sustainable growth (Pavione & Pezzetti, 2016). Cycle tourism contributes to the development of underdeveloped regions by making them "tourist attractions" (Rejon-Guardia et al., 2018). Locals benefit from increased business opportunities in the services sector, rural and remote economic development, and promotion of the areas (Brščić & Lovrečić, 2019). Community-based cycle tourism development is necessary. To promote cycle tourism, the community needs to comprehend its advantages. Entrepreneurship helps develop cycle tourism infrastructure and related goods and services (Pavione & Pezzetti, 2016). Rail trails promote entrepreneurship. Rail trails take tourists to off-the-beaten-path places. Because of this, buildings like old train stations can be turned into cafes, snack bars, or convenience stores offering opportunities for heritage revitalization. Trailside campgrounds can also be opened. Rail trails help rural and isolated communities' economic and social growth (Beeton, 2010).

Winter tourism hotels and sports facilities can be used for cycle tourism during the off-season. Teams of road cyclists who want to train at high altitude and teams of mountain bikers can go to ski resorts in the spring, fall, and summer. Ski lifts are great infrastructure that downhill mountain bikers can use when the ski season is over. Sustainable development also includes bike route integration across cities and borders. EuroVelo, the pan-European cycling network, is a great example of a cross-border bike route network in Europe (Stoffelen, 2018).

Sustainable growth requires tailoring growth to each country and region. Denmark, the Netherlands, Mallorca, and the United Kingdom invested in cycling infrastructure and services to attract cycle tourists, while France, Italy, and Spain benefited from their cycling heritage, landmarks, and popular cycling events (Weed et al., 2014). Cycling promotes sustainable transportation for tourists and locals. Investments in cycle tourism infrastructure improve public health, mobility, and recreation. Cycle tourism also inspires the development of cycling communities (Pavione & Pezzetti, 2016). When more local people use the infrastructure, the local community benefits in terms of environmental and public health (Pucher & Buehler, 2016; Goel et al., 2021). Inbound and domestic cycle tourism diversifies the tourism portfolio. Cycle tourism may encourage locals to take holidays at home more often, reducing their carbon footprint and boosting the local economy (Bakogiannis et al., 2020).

EU mechanisms for cycle tourism development

The UNWTO believes that cycle tourism satisfies a number of criteria for sustainable tourism development, including the best use of natural resources, adherence to destination authenticity and culture, socioeconomic benefits, partnerships between stakeholders, and high levels of visitor satisfaction. The European Parliament appears to concur, recognizing cycling tourism as a sustainable and ecologically favorable form of tourism (Simonsen & Jorgenson, 1998). Cycle tourism promotes rural development and helps less developed areas (Bakogiannis et al., 2020). It is regarded as a lucrative global market and is expanding substantially (Rejon-Guardia et al., 2018). It is estimated that by the year 2020, the entire continent of Europe will generate a total of 15.9 billion euros in revenue from cycle tourism (Weed et al., 2014). European plans and ideas for sustainable development have taken the growth of cycling routes and cycle tourism into account (Faulks et al., 2007). Planning and private–public stakeholder synergy are crucial for cycle tourism development. The operator network, path and route builders, promoters and providers of services, information, logistics, lodging, and others must be involved (Pavione & Pezzetti, 2016).

According to the United Nations, cycling aligns with 11 of the 17 Sustainable Development Goals (SDGs):

- Goal 1: No poverty: Cycling provides affordable transportation to education, jobs, markets, and community activities, reducing poverty.
- Goal 2: Zero hunger: Cycling can provide year-round access to food for those without other means of transportation.
- Goal 3: Good health and well-being: Cycling promotes healthy, non-polluting lifestyles, reducing heart disease and road accidents.
- Goal 5: Gender equality: Cycling gives women and girls access to water, schools, markets, and jobs that they cannot reach by walking or driving.
- Goal 7: Affordable and clean energy: E-velomobility and human power make cycling more energy efficient.
- Goal 8: Decent work and economic growth: The cycling industry creates more jobs per turnover than any other transport sector.

- Goal 9: Industry, innovation, and infrastructure: Cycling more often helps governments build resilient infrastructure and sustainable transport systems.
- Goal 11: Sustainable cities and communities: Cycling makes cities inclusive, safe, resilient, and sustainable.
- Goal 12: Responsible consumption and production: Bicycling allows people and goods to move around sustainably, matching the diversity and scale of regional and local economies.
- Goal 13: Climate action: The bicycle symbolizes reducing carbon emissions, allowing for immediate climate action.
- Goal 17: Partnerships for the goals: The cycling movement, civil society organizations, and experts advocate for public–private and civil society partnerships to promote cycling and develop environmentally friendly cycling technologies and policies worldwide.

(United Nations, 2023)

To bring transportation systems into compliance with the SDGs of the United Nations (UN), very significant changes are required. One of the main goals of public policy in the European Union (EU) has been and still is to cut down on local air pollution, accidents, and traffic jams. Promotion of walking and cycling, in combination with reliable and greener public transport, has been the main means by which the EU promotes to align its transportation systems with the SDGs (Gössling et al., 2019). The EU has been actively supporting cycling in a small but growing way. For instance, the Interreg program aids in funding cross-border and international cycling initiatives. This EU financing program funds projects that complement the EuroVelo initiative of the European Cyclists' Federation (ECF). The ECF aims to develop and integrate long-distance cycling routes across Europe. The EU funds the construction of bicycle infrastructure in undeveloped regions as well as the missing bike route linkages between nations. Just as national governments do this within each country, the EU also promotes bicycle research and the sharing of knowledge on best practices across EU nations (Pucher & Buehler, 2008).

EuroVelo is the largest EU cycle tourism initiative. It is a sustainable trans-European network of European bicycle routes. It has 19 long-distance bicycle routes, totaling 93,000 km (Eurovelo, 2023). The ECF manages the network and ensures European routes meet high design, signage, and marketing standards. EuroVelo also helps cyclists connect with trams, trains, buses, and ferries, including secure bike parking and/or bike carriage. EuroVelo expects 60 million visits to generate €7 billion in direct income (Weston et al., 2012). By centralizing all EuroVelo routes and information, the platform can boost cycle tourism at a pan-European level. Standardizing development methodologies for signage, paths, and lanes can save money on cycling infrastructure research and development. Thus, instead of researching cycling infrastructure methods, the economic resources can be used directly toward building the cycling infrastructure. EU programs promoting cycle tourism, sustainable tourism, and sustainable mobility developed and expanded most EuroVelo routes.

Several projects promoted cycle tourism. Most were funded by Interreg, some by COSME, and one by the EU Lifelong Learning Program (LLP). Programs have different objectives and methods. COSME reduces government regulation and creates a business-friendly environment to help SMEs grow. Interreg, on the other hand, promotes international cooperation and solves multi-sector issues in environment, research, education, transportation, and sustainable energy. The LLP also promotes collaboration, mobility, and exchange within education and training systems to raise their global standards and aims to help the EU become a knowledge-based, sustainable economy with social cohesion.

All three EU programs have had different objectives, aims, and methodologies and have fostered collaboration across countries and regions in different ways. COSME, Interreg, and the LLP are all similar in that they are all programs that were started and are supported by the EU. Additionally, their primary goals are to encourage cooperation and integration among member states. In addition, the goals of all three programs are to make a contribution to the EU's long-term economic growth and development, with an emphasis on fostering social cohesion and increasing the number and quality of employment opportunities.

To further coordinate cycling-specific projects, the European Commission has developed a detailed manual for cycling projects in the EU, titled Guidance for Cycling Projects in the EU. The guidance for cycling projects in the EU is intended for practitioners working for city authorities and other stakeholders involved in cycling. It includes city case studies, best practices, and relevant EC-funded projects in the cycling industry (European Commission, n.d.).

EU programs promoting bicycle tourism

In the period between 2010 and 2023, the EU financed a plethora of projects that were meant to promote and grow cycle tourism. These projects included the development of cycling packages and services, the promotion of sustainable tourist practices, the building of digital infrastructure for the promotion and information of cycle tourism, the creation of bike routes, and the upgrading of existing infrastructure. Moreover, the projects also aimed to improve collaboration across international borders and encourage economic development in the implementation areas.

Table 5.1 provides an overview of popular EU-funded projects for cycle tourism promotion and development.

The EU projects that were mentioned tried to improve the growth of sustainable tourism by promoting cycling as a way to move around in a sustainable way and cycle tourism as a way to travel in a sustainable way. This was done in order to meet the objectives of the projects. The projects concentrated on a variety of areas of cycle tourism, including policy instruments, co-creation, co-management, destination governance, non-formal learning, shared and integrated solutions, and economic recovery, among other topics. The initiatives aimed to foster sustainable tourism in a variety of places, including both urban and rural areas, and to stimulate the use of cycling for both shorter and longer distances, with the goal of lowering the carbon footprint and the seasonality of tourism. Despite the fact that every

Table 5.1 Overview of EU-funded cycle tourism projects

Project name/description	Funding body	Budget
Cycling for development, growth, and quality of life in European regions (EU-CYCLE)	Interreg Europe	€1.06M
European Cycle Tourism Educational Program EU Bike Project	European Union Lifelong Learning programme, between 2014 and 2016	€529.5K
Amazon of Europe Bike Trail (AoE Bike Trail)	Interreg Danube	€3.17M
European network for the promotion of cycle tourism in natural areas (ECO-CICLE)	Interreg Europe	€1.35M
Sharing best practices and experience on data collecting and processing and involvement of users in order to improve planning of cycling and walking as modes of transport in urban and functional urban areas (CYCLEWALK)	Interreg Europe	€1.59M
The Rhine Cycle Route (DEMARRAGE)	Interreg IV B Northwest Europe	€2.3M
Atlantic on Bike (AoB)	Interreg Atlantic Area	€4.58M
MEDiteranean CYcle route for sustainable coastal TOURism (MEDCYCLETOUR)	Interreg Mediterranean	€2.5M
Encouraging cycling mobility in rural areas (BICIMUGI)	Interreg POCTEFA	€4.4M
Pilgrims Routes EuroVelo 3 – COSME	COSME	€363K
Promotion and development of the Baltic Sea Cycle Route (Route No. 10) in Denmark, Germany, Lithuania, Poland, and Sweden (Biking South Baltic)	Interreg South Baltic	€988K
Transdanube.Pearls – Network for Sustainable Mobility along the Danube (TRANSDANUBE.PEARLS)	Interreg Danube Transnational Programme	€2.93M
EuroVelo 5 – Via Romea (Francigena)	COSME	€250K
Silver Cyclists	COSME	Unknown
EuroVelo 13 – Iron Curtain Trail	Interreg Austria – Czech Republic	€1.4M

project had its own unique goals and objectives, they all had the same overarching vision: to improve the quality of life in the communities in which they were located by fostering more sustainable and responsible forms of tourism, to enhance the natural and cultural heritage of the area, and to promote conservation efforts. These EU projects not only underlined the significance of cross-border collaboration and the exchange of information in the process of accomplishing this objective, but they also illustrated the potential of cycling tourism as a sustainable alternative to traditional forms of tourism.

Many bike trails that span international borders have been built, and the bulk of them were made possible thanks to funding from the Interreg program. Nevertheless, in spite of the possibility of positive quantitative indicators, the lack of coordination between the projects and the stakeholders has led to their questionable longevity, uncertain tourism potential, and the absence of a strategic vision regarding how to combine these local efforts in order to achieve benefits for the entire cross-border region. It is of the utmost importance to create a forum for two-way communication and cooperation across international borders among the many parties participating in these initiatives. As a result, not only will this serve to encourage regional growth and tourism, but it will also help secure the long-term viability and efficiency of the bike routes (Stoffelen, 2018). Interreg transnational projects have often been criticized for being unsustainable. Supranational organization-backed short-term projects may hinder the development of a sustainable cross-border tourism industry. The project and its funding usually end, and its results become obsolete (Shepherd & Ioannides, 2020). Interreg projects are also bureaucratic and inflexible. Project management, dissemination, travel for transnational meetings, expenditure evaluation, and procurement of external consultants, services, equipment, and infrastructure consume a large portion of the budget due to their bureaucratic and inflexible nature. Thus, fewer resources and efforts go to the actual development of infrastructure (McMaster et al., 2019; Van et al., 2020).

Despite the aforementioned criticism, EU programs have served greatly for the development of cycle tourism in Europe. The implementation of EU projects for cycle tourism has proven to be a best practice in sustainable tourism development as a result of the organized and holistic approach taken. The reason behind the successful and sustainable implementation of EU projects for cycling tourism is the involvement of the ECF. Projects related to cycle tourism development were implemented under the umbrella of the EuroVelo initiative, and the main outputs that were produced by the various projects were directly implemented into the pan-European cycling network. The ECF played a crucial role in coordinating and providing technical expertise to the projects' partners, ensuring that the cycling infrastructure and services are based on EuroVelo methodologies and standards. As a result, EuroVelo has arguably become the most extensive and well-developed cycle tourism network in the world with the help of EU project implementation. In conclusion, the development of transnational cycling destinations with the help of EU funds has been shown to be a truly sustainable approach.

In conclusion, the EU's support for the growth of cycle tourism has shown to be an effective and sustainable way to promote sustainable tourism and mobility. The achievement of positive outcomes served as the proof of this. The projects that have been reported to have been carried out by the EU have concentrated on various aspects of bicycle tourism. The overarching objective of these projects is to encourage environmentally responsible tourism in both urban and rural areas, and to promote the use of cycling as a means of reducing carbon footprints and the seasonality of tourism. However, there is a lack of coordination and a strategic vision regarding how to combine these local efforts in order to achieve benefits for the entire EU region, which has raised concerns about the longevity of the projects and the potential for tourism in the region. In spite of these criticisms, the involvement of the ECF has played a crucial role in ensuring the implementation of EuroVelo methodologies and standards, which have resulted in what is arguably the most extensive and well-developed network of cycle tourism anywhere in the world. The European Commission has contributed to the coordination of cycling-related projects with the specific guidelines it released. In general, EU programs have been of great assistance in the expansion of cycle tourism in Europe. These programs have provided a model for the most effective method of promoting sustainable tourism development through an approach that is both organized and comprehensive.

References

Andreev, H., & Zopiatis, A. (2022). *Cycle tourist characteristics*. In *Encyclopedia of tourism management and marketing* (pp. 756–758). Cheltenham Glos, UK: Edward Elgar Publishing.

Bakogiannis, E., Vlastos, T., Athanasopoulos, K., Christodoulopoulou, G., Karolemeas, C., Kyriakidis, C., Noutsou, M. S., Papagerasimou-Klironomou, T., Siti, M., Stroumpou, I., & Vassi, A. (2020). Development of a cycle-tourism strategy in Greece based on the preferences of potential cycle-tourists. *Sustainability*, 12(6), 2415.

Beeton, S. (2010). Regional community entrepreneurship through tourism: the case of Victoria's rail trails. *International Journal of Innovation and Regional Development*, 2(1–2), 128–148.

Brščić, K., & Lovrečić, K. (2019, July). *How to plan the development of cycle tourism? – Example of Istria county*. TISC-Tourism International Scientific Conference Vrnjačka Banja, 4(2), 603–620.

Buning, R. J., Cole, Z., & Lamont, M. (2019). A case study of the US mountain bike tourism market. *Journal of Vacation Marketing*, 25(4), 515–527.

Buning, R. J., & Gibson, H. J. (2016). The role of travel conditions in cycling tourism: Implications for destination and event management. *Journal of Sport & Tourism*, 20(3–4), 175–193.

Coghlan, A. (2012). An autoethnographic account of a cycling charity challenge event: Exploring manifest and latent aspects of the experience. *Journal of Sport & Tourism*, 17(2), 105–124.

Derom, I., & Ramshaw, G. (2016). Leveraging sport heritage to promote tourism destinations: The case of the Tour of Flanders Cyclo event. *Journal of Sport & Tourism*, 20(3–4), 263–283.

Dickinson, J., & Lumsdon, L. (2010). Cycling and tourism. In R. Sharpley (Ed.), *Slow travel and tourism* (pp. 135–146). London: Earthscan.

ENAT. (2016). European network for accessible tourism. *Silver Cyclists: Using EuroVelo, the European Cycle Route Network, to Attract More Seniors to Cycling Tourism in the Low and Medium Season*. Retrieved from www.accessibletourism.org/?i=enat. en.enat_projects_and_good_practices.1931

European Commission. (n.d.). *Guidance for cycling projects in the EU*. Retrieved from https://transport.ec.europa.eu/transport-themes/clean-transport-urban-transport/cycling/guidance-cycling-projects-eu_en

Eurovelo. (2021). *Promoting cycling tourism in natural areas*. Final Event of the ECO-CICLE Project at the EuroVelo & Cycling Tourism Conference. Retrieved from https://pro.eurovelo.com/news/2021-11-24_promoting-cycling-tourism-in-natural-areas-final-session-of-the-eco-cicle-project-at-the-eurovelo-cycling-tourism-conference/

Eurovelo. (2023). *EuroVelo data hub*. Retrieved from https://pro.eurovelo.com/projects/eurovelo-data-hub/

Eurovelo – Cosme – EuroVelo 3. (n.d.). Retrieved from https://pro.eurovelo.com/projects/2019-09-10_cosme-eurovelo

Eurovelo – Cosme – EuroVelo 5. (n.d.). *VIA ROMEA EuroVelo 5 – Via romea (Francigena)*. Retrieved from www.velo-territoires.org/schemas-itineraires/schema-europeen-eurovelo/eurovelo-5/

Eurovelo – Demarage – EuroVelo 15. (n.d.). Retrieved from https://pro.eurovelo.com/projects/2019-09-10_demarrage-eurovelo-15-rhine-route

Faulks, P., Ritchie, B., & Dodd, J. (2008, December). *Bicycle tourism as an opportunity for re-creation and restoration? Investigating the motivations of bike ride participants.* New Zealand Tourism and Hospitality Research Conference, Canterbury: New Zealand, 1–27.

Faulks, P., Ritchie, B., & Fluker, M. (2007). *Cycle tourism in Australia: An investigation into its size and scope.* Gold Coast: Sustainable Tourism Cooperative Research Centre.

Gibson, H., & Chang, S. (2012). Cycling in mid and later life: Involvement and benefits sought from a bicycle tour. *Journal of Leisure Research*, 44(1), 23–51.

Goel, R., Goodman, A., Aldred, R., Nakamura, R., Tatah, L., Garcia, L. M. T., . . . & Woodcock, J. (2021). Cycling behaviour in 17 countries across 6 continents: levels of cycling, who cycles, for what purpose, and how far? *Transport Reviews*, 1–24.

Gössling, S., Choi, A., Dekker, K., & Metzler, D. (2019). The social cost of automobility, cycling and walking in the European Union. *Ecological Economics*, 158, 65–74.

Interreg. (n.d.). *About Interreg.* Retrieved from https://interreg.eu/about-interreg//

Interreg Atlantic Area: Atlantic On Bike. (n.d.). *The EuroVelo 1, a unique cycling-tourism destination for a green growth.* Retrieved from https://atlanticarea.eu/project/32

Interreg Danube. (n.d.). Transdanube.Pearls. *Transdanube.Pearls – Network for Sustainable Mobility along the Danube.* Retrieved from www.interreg-danube.eu/approved-projects/transdanube-pearls/

Interreg Danube – AoE. (n.d.). *Amazon of Europe bike trail.* Retrieved from www.interreg-danube.eu/approved-projects/amazon-of-europe-bike-trail

Interreg Europe: Cyclewalk. (n.d.). *Sharing best practices and experience on data collecting and processing and involvement of users in order to improve planning of cycling and walking as modes of transport in urban and functional urban areas.* Retrieved from https://projects2014-2020.interregeurope.eu/cyclewalk/

Interreg Europe: Eu Cycle. (n.d.). *Cycling for development, growth and quality of life in European regions.* Retrieved from https://projects2014-2020.interregeurope.eu/eucycle/

Interreg Mediterranean – Medcycletour. (n.d.). *MEDiteranean CYcle route for sustainable coastal TOURism.* Retrieved from https://medcycletour.interreg-med.eu/

Interreg Pogtefa – Bicimugi. (n.d.). *Encouraging cycling mobility in rural areas.* Retrieved from https://en.bicimugi.eu/

Interreg South Baltic. (n.d.). *Biking South Baltic! – Promotion and development of the Baltic Sea cycle route (route no. 10) in Denmark, Germany, Lithuania, Poland and Sweden.* Retrieved from https://southbaltic.eu/-/biking-south-baltic-promotion-and-development-of-the-baltic-sea-cycle-route-route-no-10-in-denmark-germany-lithuania-poland-and-sweden/

Iron Curtain Trail. (n.d.). *Old borders newly experienced at the Iron Curtain Trail.* Retrieved from www.ict13.eu/iron-curtain/about-the-project

Jurišić, M., Jerkunica, A., & Jeli, M. (2018). *Perspectives of cycle tourism in the case of Split-Dalmatia County.* 2. Kongresa Sportskog Turizma, Makarska, Hrvatska, 11. i 12. studeni 2016. godine. Zbornik odabranih znanstvenih radova, 50–60.

Katelieva Platzer, M., Mitrofanenko, T., & Palhau Martins, C. (2017). *Cooperative development of cycle tourism in Europe-EuBike project.*

Lamont, M. (2009). Reinventing the wheel: A definitional discussion of bicycle tourism. *Journal of Sport & Tourism*, 14(1), 5–23.

Lee, C. F., & Huang, H. I. (2014). The attractiveness of Taiwan as a bicycle tourism destination: A supply-side approach. *Asia Pacific Journal of Tourism Research*, 19(3), 273–299.

Legan, N. (2017, October 19). *Gravel cycling: The complete guide to gravel racing and adventure bikepacking.* New York, Netherlands: VeloPress, Illustrated edition.

Lőrincz, K., Banász, Z., & Csapó, J. (2020). Customer involvement in sustainable tourism planning at Lake Balaton, Hungary – Analysis of the consumer preferences of the active cycling tourists. *Sustainability*, 12(12), 5174.

McMaster, I., Wergles, N., & Vironen, H. (2019). Results orientation. *European Structural and Investment Funds Journal*, 7(1), 2–8.

Pavione, E., & Pezzetti, R. (2016). Cycle tourism as a form of sustainable tourism: The importance of a policy for its enhancement. *Opportunities and Risks in the Contemporary Business Environment*, 997.

Pucher, J., & Buehler, R. (2008). Making cycling irresistible: Lessons from the Netherlands, Denmark and Germany. *Transport Reviews*, 28(4), 495–528.

Pucher, J., & Buehler, R. (2016). Safer cycling through improved infrastructure. *American Journal of Public Health*, 106(12), 2089–2091.

Rejón-Guardia, F., García-Sastre, M. A., & Alemany-Hormaeche, M. (2018). Motivation-based behaviour and latent class segmentation of cycling tourists: A study of the Balearic Islands. *Tourism Economics*, 24(2), 204–217

Shepherd, J., & Ioannides, D. (2020). Useful funds, disappointing framework: Tourism stakeholder experiences of INTERREG. *Scandinavian Journal of Hospitality and Tourism*, 20(5), 485–502.

Shipway, R., King, K., Lee, I. S., & Brown, G. (2016). Understanding cycle tourism experiences at the Tour Down Under. *Journal of Sport & Tourism*, 20(1), 21–39.

Simonsen, P., & Jorgenson, B. (1998). *Cycle tourism: An economic and environmental sustainable form of tourism*. Bronholm, Denmark: Unit of Tourism Research, Research Centre of Bornholm.

Stoffelen, A. (2018). Tourism trails as tools for cross-border integration: A best practice case study of the Vennbahn cycling route. *Annals of Tourism Research*, 73, 91–102.

United Nations. (2023). *Cycling and Sustainable Development Goals*. Retrieved from https://unric.org/en/sustainable-development-goals-cycling/

Van Den Broek, J., Rutten, R., & Benneworth, P. (2020). Innovation and SMEs in Interreg policy: Too early to move beyond bike lanes? *Policy Studies*, 41(1), 1–22.

Varnajot, A. (2020). The making of the Tour de France cycling race as a tourist attraction. *World Leisure Journal*, 62(3), 272–290.

Weed, M., Bull, C., Brown, M., Dowse, S., Lovell, J., Mansfield, L., & Wellard, I. (2014). A systematic review and meta-analyses of the potential local economic impact of tourism and leisure cycling and the development of an evidence-based market segmentation. *Tourism Review International*, 18(1), 37–55.

Weston, R., Davies, N., Peeters, P. M., Eijgelaar, E., Lumsdon, L., McGrath, P., & Piket, P. C. (2012). The European cycle route network EuroVelo: Challenges and opportunities for sustainable tourism. *Update of the 2009 Study*.

6 The use of football stadia as venues for meetings and conferences

The case of the UK

Rob Davidson and Andrew Kirby

Demand-side perspectives

Since the 1990s there has been a movement, particularly within the European conference and meetings industry, for organisers of events to use unusual, 'non-traditional' or 'unique' venues such as museums, theatres, sports facilities and other visitor attractions in place of the more traditional meeting venues such as city centre hotels and conference centres. From the perspective of meeting planners – and corporate meeting planners in particular – offering delegates and attendees the opportunity to visit these new, unexplored types of venues can be powerful measures of persuasion in generating event attendance and motivating participants. Consequently, a growing number of planners are seeking venues that create a unique and memorable experience for their attendees (Phillips & Geddie, 2005; Rogers & Davidson, 2016; Leask & Hood, 2001; Lee & Kim, 2015; Dowson & Wilson, 2023).

Several authors have identified the widespread trend towards smaller and shorter, one-day meetings not requiring overnight accommodation as being an important demand-side factor contributing to the growth in the use of unusual venues (Rogers & Davidson, 2016; Maritz, 2012). Respondents to the Maritz survey added a number of other motivations for using non-traditional venues, including 'Meetings are looking for options other than "cookie-cutter" type of venues' and 'Attendance increases at unique venues, and delegates are more engaged, interested and therefore more apt to learn and gain from the experience'. Leask and Hood (2001) and Rogers and Davidson (2016) concur, emphasising the novelty, memorability and uniqueness of the participants' experiences while attending events hosted in unusual venues. As two additional advantages of using unusual venues to host conferences, Rogers and Davidson (2016) suggest that the venue itself can be a topic of conversation among participants – a useful icebreaker in networking sessions, for example, and the venue may have a link with the theme of the conference – for instance, using meeting space in an aquarium to host a conference on marine biology.

Sports facilities have the potential to offer particular advantages, as noted by Lee and Kim (2015): 'A sport facility has a unique value, particularly if the team in the venue has a strong brand. This would only enhance the site as a desirable event/ meeting space, by merging the team's image with the event theme'.

DOI: 10.4324/9781003384786-6

But a number of authors have also noted that the use of unusual venues can have their own particular disadvantages that are not normally shared by purpose-built venues. These include the lack of overnight accommodation, in-house catering facilities and audio-visual technology; restrictions due to licensing laws, Listed Building Regulations and the requirement not to interfere with the public's enjoyment of the facility's primary purpose; the lack of meetings industry knowledge on the part of the venue's staff and the peripheral location of some unusual venues (Leask & Hood, 2001; Rogers & Davidson, 2016; Lee & Kim, 2015). In recent years, however, as will be discussed on this chapter, many of these disadvantages have diminished.

Supply-side perspectives

From the perspective of the facilities themselves, developing a secondary function as a meetings venue has been welcomed by many as a means of creating an additional revenue stream by making use of their unused capacity. In the UK, this trend coincided with a time when government subsidies in the arts and culture were being cut back significantly. As noted by Leask and Hood (2001), 'Reduced government funding has encouraged many of these venues to diversify their revenue generation activities, with conference activity providing lucrative and innovative opportunities'.

In recent decades, many sports venues worldwide, including football stadia, have adopted a strategy of diversifying their offer in order to raise additional financial income streams generated from non-matchday activities and sustain year-round commercial activities (Marr, 2011). Edensor et al. (2021) identified four different dimensions of football stadium experiences that are being offered by a growing number of facilities: stadia as museums, stadia as pilgrimage sites, stadia as restaurants, and stadia as event venues. There is a direct link between this type of expansion and diversification in the use of football stadia and the United Nations (UN) Sustainable Development Goal (SDG) 8, which includes the objectives 'to promote policies to support job creation and growing enterprises' and 'to improve resource efficiency in consumption and production'. For example, opening football stadia for the hosting of conferences on non-matchdays is a clear demonstration of using a valuable resource more efficiently, in terms of both consumption and production.

However, a challenge for many football stadia has been to dismiss previous outdated perceptions of poor-quality, old-fashioned facilities and catering and to provide clear evidence of the improvements and renovations that have taken place in such venues, many of which now offer a serious and cost-effective solution to the events industry. Wealthier British football clubs in particular have invested in expensive, innovative stadia design, enhanced comfort, corporate boxes, hospitality provision and accessibility, and, crucially, in commercial areas beyond football and the matchday experience (Paramio et al., 2008; Sheard, 2005).

As stadia update and refurbish their facilities to create more modern and efficient building, there has been a strong focus on sustainable initiatives, such as improving insulation, efficient glazing (to minimise heat loss), reducing energy use and water consumption, installing efficient LED lighting and exploring options for solar panel energy generation. Many venues also engage in carbon off-setting with the planting of trees and flowers to balance the carbon generated by their operations (www.stadiumexperience.com). New-built stadia, in particular, have used the opportunity to respect the principles embodied in UN SDG 12: responsible consumption and production. For example, upon its opening, the new Tottenham Hotspur FC Stadium topped the 2020 Premier League Sustainability Table, taking into account key factors such as renewable energy use, carbon emissions, plastic use and water saving. The stadium is recognised as 100% certified on renewable energy use. As a consequence, carbon dioxide emissions from the stadium are around 50% less than that from a stadium built 10 years ago (ibid.)

As a result of such investments, many UK football stadia are now able to successfully position themselves as effective and sustainable venues for meetings. The design and construction of several high-profile new-built football stadia in England within the Premier League, including the Emirates Stadium (home of Arsenal FC), the new Tottenham Hotspur FC Stadium, and the Amex Community Stadium (home of Brighton and Hove Albion FC), have all seen the inclusion of non-matchday usage as a key strategic income stream directly affecting the business and finances of the clubs themselves. The promotion of such stadia to the European and international meetings market allows the clubs to potentially maximise the use of their assets and generate significant income during times when matches are not being held and large corporate hospitality rooms are normally empty.

The experience

Naturally, the distinctive experience offered to those attending events at high-profile football stadia is a key unique selling point that is heavily promoted by those responsible for attracting events and conference business to such locations. Many sporting fans are strongly attracted to business events and corporate meetings at these venues, and they appreciate the opportunity to see behind the scenes at famous stadia and create personal memories in grounds seen on national and international sporting media each weekend. Venues have sought to offer as many creative means as possible to enhance the meeting attendees' overall experience, such as personalised stadium tours, additional dedicated time in club merchandising and retail outlets, informal talks by ex-players on their careers and visits to club museums. With minimal additional cost implications to stadia themselves, these add-ons are often highly successful as strategic marketing tools and can play a significant part in the final decision-making process of national and international event and conference organisers in their venue selection as they offer additional perceived value to their attendees.

Furthermore, particularly for international conferences, many UK football clubs offer the possibility of combining such events with attendance at either midweek or Saturday afternoon matches, allowing attendees to experience a game first hand, often for the first time. Given the high level of demand for match tickets, this particular added-value benefit can act as an additional incentivising factor for conference organisers.

Accessibility

The location of football stadia in urban centres across the UK provides logistical benefits to event organisers seeking central or regional event venues with the power to attract attendees from large urban destinations and cities. Due to their matchday requirements with many thousands of attendees, football stadia often have excellent transport links from major population hubs including strong public transport networks and easy motorway and road access. Given such demands on matchdays, many of these venues also provide multiple car parking options, often for free on non-matchdays, offering an advantage over those city-centre event venues where car parking can often be at a premium and excessive cost. For meetings attendees travelling to the venue by car, installation of electric vehicle charging points close to football stadia has also grown substantially in recent years. These can be found, for example, at the Easter Road Stadium in Edinburgh. Such chargers are often free to use for non-matchday event attendees.

Accommodation

While lack of accommodation has been identified as a disadvantage for many unusual venues, the offer of budget-style hotels located close to many football venues may be regarded as an additional asset for facilities that are able to attract residential business events. There are many examples of high-profile football clubs within the English Premier League – for example Manchester United FC, Manchester City FC, Southampton FC and Chelsea FC – where hotels and other budget accommodation providers have constructed properties in or close to these stadia.

Catering

Catering is a key area of operations where football stadia have invested significant time and finance over recent decades to dismiss the previous 'Pie and a Pint' reputation held by many, and which could often be a significantly negative factor for such venues seeking to attract corporate and association event attendance. The addition of professional chefs, the introduction of new creative menu offerings and purpose-built new restaurant facilities have all contributed to dismissing the poor image of football stadia catering and now feature strongly in the marketing of such venues in order to inform decision-makers in the meetings industry about the key advances made in recent years.

Many of those advances have been in harmony with the principles contained in UN SDG 12, and a growing number of stadia are following sustainable practices as they apply to their food and beverage offer, including for meetings and events. For example, Watford FC's 'Mindful Meeting' events catering package includes such features as Eco LEAP labelling of all dishes; a higher percentage of plant-based dishes; British-sourced fruit and vegetables, rather than air freight produce; unlimited tea and coffee served in reusable crockery; and no single-use plastic (www.watfordfcevents.com).

The importance of attracting event organisers in person to venues to witness and experience such improvements has also been a key success factor in raising the profile of sports venues and increasing the amount of business secured to football clubs in recent times. Even some lower-level football clubs in England have created partnerships with high-profile celebrity chefs to host and run branded restaurants in their stadia, both to increase the profile of the club itself and to boost their culinary offer, for example, chef Marco Pierre White's partnership with the Milton Keynes Dons, a League One UK football club.

Technology

Another important requirement for professional conference organisers and their participants is reliable and cost-effective Internet access both for use within daytime meeting attendance and for attendees staying overnight. Internet access is an area in which football stadia have invested heavily, to enable them to attract both small- and large-scale events of different types to their venues. Wembley Stadium, for example, the UK's national football stadium, has an ongoing partnership with the communications giant EE and actively promotes its reliable Internet access throughout the stadium, offering connectivity on matchdays with tens of thousands of fans and also on non-matchdays for corporate and professional association events.

Marketing

Football stadia in the UK have adopted the use of many of the marketing channels employed by more traditional venues such as conference centres and hotel chains. For example, attendance at domestic and international meetings industry trade shows such as IMEX, International Confex and IBTM World by clubs such as Manchester United FC, Manchester City FC and Liverpool FC provide these venues with the opportunity to communicate with a wide range of international event and conference planners and to increase the awareness of the new improved facilities at their locations.

Many stadia are assisted in their marketing by Stadium Experience (https://stadiumexperience.com), a not-for-profit marketing collaboration of around 40 football, rugby and cricket club venues across the UK that work together to promote stadia as effective venues for conferences, meetings, exhibitions or even parties. Sharing of information, ideas and best practice among stadia is a key objective of Stadium Experience. Their activities include the following:

Members Meetings: All members are invited to share best practices and ideas, bring relevant topics to the table and meet potential suppliers.
Public Catering Meetings: Give clubs' public catering & kiosk managers the opportunity to share best practices and ideas that they use in their meeting spaces.
Swap Shop: A system of visits of personnel between football clubs, enabling them to connect with other clubs, share ideas and provide feedback to aid improvements.
Agent Networking: Key events agents are invited to attend Members Meetings to present and network with stadia personnel. In addition, Stadium Experience arranges regular meetings for football clubs event personnel to visit agents and

present details of their venues. Recent collaborations have included meetings with Calders, Edge Venues, HelmsBriscoe, Jigsaw Events, KTS Events, Liz Hobbs Group, Meetings Industry Association, Off Limits, Space2, The Conference Group, Venue Directory, Venue Finder and BCD.

Exhibitions: Stadium Experience offers its members the opportunity to stand-share and have a presence at meetings industry exhibitions at a reduced price. Recent examples of such exhibitions include the Corporate Hospitality Show Leeds, the Corporate Hospitality Show Birmingham, the Inspirational Venue Roadshow, the Meetings Show, 20/20 Speed Networking Events, Event Organisers Summit and the Event Agency Forum.

The Stadium Events & Hospitality Awards: This annual award programme was founded in 2005 by members of Stadium Experience to recognise the very best practice in UK football stadia venues. In recent years, the range of awards has expanded to include the recognition of excellence in the non-matchday aspects of stadia management and marketing, such as conferences – for example the Non-Matchday Sales Team of the Year Award.

Conclusion

Overall, it is clear that football stadia in the UK and elsewhere have made significant advances in recent years to position themselves as a cost-effective and sustainable venue option within the modern-day conference and events marketplace. Organisers and delegates are increasingly attracted by the significant new infrastructure offering improved catering, added-value activities and knowledgeable staff able to deliver high-class services for the meetings market. The unique atmosphere and 'wow factor' generated by sports venues, even on non-matchdays, offer clients the opportunity to add value to their events and differentiate their offering to national and international events attendees. With a multitude of traditional and non-traditional venues for meeting planners to choose from, football stadia are asserting themselves as serious and sustainable meeting spaces for football fans and non-fans alike.

Several questions remain unanswered however, and further research into the demand-side factors in particular is required. For instance, a sound case can be made for an investigation into whether participating in a meeting held in a football stadium holds as much attraction for women as it does for men. It might be reasonable to assume that in this age of growing visibility of women's football (itself linked to UN SDG 5, to promote gender equality and empower all women and girls), the use of stadia as meeting venues would hold more appeal for women. But this hypothesis remains to be properly tested by research.

References

Dowson, R., & Wilson, K. (2023). Alternative venues for business events. In *The Routledge handbook of business events* (pp. 117–129). Routledge.

Edensor, T., Millington, S., Steadman, C., & Taecharungroj, V. (2021). Towards a comprehensive understanding of football stadium tourism. *Journal of Sport & Tourism*, 25(3), 217–235.

Leask, A., & Hood, G. L. (2001, June). Unusual venues as conference facilities: current and future management issues. In *Journal of convention & exhibition management* (Vol. 2, Issue 4, pp. 37–63). Abingdon, UK: Taylor & Francis Group.

Lee, S., Parrish, C., & Kim, J.-H. (2015). Sports stadiums as meeting and corporate/social event venues: A perspective from meeting/event planners and sport facility administrators. *Journal of Quality Assurance in Hospitality & Tourism*, 16(2), 164–180.

Maritz. (2012). The future of meeting venues white paper. *Maritz Research*.

Marr, S. (2011). Applying "work process knowledge" to visitor attractions venues. *International Journal of Event and Festival Management*, 2(2), 151–169.

Paramio, J. L., Buraimo, B., & Campos, C. (2008). From modern to postmodern: The development of football stadia in Europe. *Sport in Society*, 11(5), 517–534.

Phillips, W., & Geddie, M. (2005). An analysis of cruise ship meetings factors influencing organization meeting planners to select cruise ships over hotels for meetings. *Journal of Convention & Event Tourism*, 7(2), 43–56.

Rogers, T., & Davidson, R. (2015). *Marketing destinations and venues for conferences, conventions and business events*. London, UK: Routledge.

Sheard, R. (2005). *The stadium: Architecture for the new global culture*. Berkeley: Periplus Editions.

www.stadiumexperience.com. *Focusing on sustainability at stadium venues*. Retrieved on 2023 from https://stadiumexperience.com/2022/07/focusing-on-sustainability-at-stadium-venues/

www.watfordfcevents.com. *Our food philosophy*. Retrieved on 2023 from www.watfordfcevents.com/food-drink/our-philosophy/

7 Designing for the beautiful game

Soccer stadia for a sustainable future

Terry Stevens

Introduction

"Even now, whenever I arrive at any football ground, or merely pass close to one when it is silent, I experience a unique alerting of the senses. The moment evokes my past in an instantaneous emotional rapport which is more certain, more secret, than memory."

(Hopcroft, 1971)

Since the Second World War, the sports stadium has become the iconic building of the modern age. Able to encapsulate identities of entire communities and nations, these structures have few parallels in contemporary society. They make a significant contribution to destination place-making and branding. They directly affect the reputation and success of soccer clubs and influence the behaviour of fans and spectators, both during a game and in their everyday lives (Filger, 1981; Fishwick, 1989; Holt, 1989; Buford, 1992; Birley, 1993; Bellos, 2002; Foer, 2010).

Stadia provide a forum for the creation of sense of place and identity – "(Their) construction, re-construction and celebration of cultural practices contributing this to the re-definition of identity" (Budka & Jacano, 2013; Sandry, 2018). As Strutt (1801) observed, "In order to form a just estimate of the character of any particular people it is absolutely necessary to investigate the sports and pastimes most prevalent amongst them." Two hundred years later, Inglis was to add stadia to this baseline. For Inglis, stadia are cultural barometers – and have been since ancient times (Inglis, 1993; Inglis, 2000). Empty or full, ultra-modern or decrepit, they offer gateways to an understanding of how not only fans but also whole cities and communities operate. Or as Inglis believes, "by their stadiums you shall know them."

Soccer stadia, especially those with the potential for multiple or dual use, and other types of major sport and cultural events, help create a vibrant destination. They are fundamental features of a destination's infrastructure, appeal, and image (Ritchie & Adair, 2004; Weed & Bull, 2004; Hinch & Higham, 2011). Soccer stadia have often been referred to as "cathedrals of sport" (Stevens, 2005; John & Sheard, 2000).

It is apparent that, despite the dramatic increase in interest about stadia over the past 20 years, it has been only recently that sustainability has featured as a topic but has rapidly become a major focus of interest. Sustainability in stadia

DOI: 10.4324/9781003384786-7

development was placed centre stage at the First World Sports Tourism Congress in 2021 (UNWTO & ACT, 2022), which concluded that "Excellence is expected in all aspects of the design and management of these facilities. It is now taken for granted that these venues will be safe, secure, and hygienic for the competing athletes, their support staff, and spectators. They must also be designed and managed with consideration for the communities that lives around these venues and integrated with the local environment, especially public transport systems."

The Congress recognised that the modern stadia and arenas are complex environments where the specific needs of the athletes and the spectators have to be met. This entails creating the right conditions for elite performance along with the safety, security, and comfort for all others – spectators, support staff, media, and those working in these venues. The viability of these facilities, often expensive to build and operate, demands quality provision of commercial services (hospitality, retail, and entertainment) along with medical, administrative, and other amenities. There is considerable pressure to ensure that these facilities can be fully used throughout the year for non-sporting events to drive revenues and viability. It has now become essential to apply innovative thinking to how stadia and arena can be developed for multiple use in a sustainable manner. This includes novel forms of visitor attractions such as museums, visitor centres, and adrenalin experiences (such as ziplines, roof walks, and rollercoaster rides on the roofs of grandstands) as part of new sustainable business models.

Finding sustainable solutions to reducing costs and generating income is fundamental. Examples range from everyday practice of recycling waste to forms of renewable energy, stadia having their own farms for food production, the collection and reuse of rainwater for irrigating grass pitches, and creating nature conservation habitats around the built facilities. Innovation is particularly important in the search for sustainable solutions. Stimulating fresh, innovative, thinking requires a working environment where anyone can promote their "good idea" – there is no hierarchy for good ideas. Collaboration and new forms of partnerships are essential to make good ideas work. Creating the delivering new ways of working can take time, but little steps lead to successful big ideas. Innovation is the key to being competitive. The world of the sustainable sports stadium is not separate from the innovation of society, the progress of society is continuous, and sports stadia have to follow this rhythm, if possible anticipating it. The Congress concluded that a stadium is not just about sport, it must reassure the community that these structures are responding to them in many ways whilst addressing the need to be economically, socially, and environmentally responsible.

In this way, it is clear that the UNWTO and others acknowledge that the next generation of stadia development must fully embrace the principles of the triple bottom line – social, economic, and environmental sustainability. These principles are as applicable to the re-imagining of The Racecourse (home to Wrexham FC, the world's oldest in-use international soccer ground in the world) by its new American owners (The Football Ground Guide, 2023), as well as the re-development of FC Barcelona's Camp Nou (Col-legi d'Arquitectses de Catalunya, 2018), and for the creation of new soccer stadia such as the Cathedral in Milan (replacing the San Siro), the Eco Park for Forest Green Rovers (in Stroud, England), or the planned new stadia for Feyenoord in Rotterdam and GNK Dinamo Zagreb (Dezeeen, 20.2.30).

The challenge to embrace the sustainable approach is significant. KPMG (2013) estimated that there were some 330 football stadia in Europe with a capacity of 20,000 or more: UK with 54, Germany with 47, Italy with 35, and Spain with 29 venues lead the rankings. KPMG state that in Europe, countries with no stadia of 20,000 seats or more are typically small and with a less competitive domestic football product. Further afield in the Middle East and Africa, there is a high volume of stadia in the largest countries or countries with a strong sports tradition. At the time KPMG delivered their report, it was estimated that 90% of European stadia with a capacity of over 20,000 (60% including the selected African and Middle Eastern countries) are more than 30 years old and may require major renovation or upgrade to stay "in sync" with the expectations of today's agendas for the fan, the community, and the SDGs. Renovation does not provide an optimum solution. New build is more likely to deliver these objectives. The hurdles to realising this are often substantial, and with the right concept and business plan, innovative stadium design, and the right team in place, these hurdles can be overcome. The 2013 KPMG *A blueprint for successful stadium development* detailed a helpful, positive, and sustainable way forward. Union of European Football Associations (UEFA's) "Strength Through Unity Strategy" and their "Environment, Social Responsibility and Corporate Governance ESG" model linked to the UEFA stadium star rating criteria now reflect many aspects of the KPMG blueprint (UEFA, 2018) and are now underpinned by UEFA providing, free of charge, consultancy to clubs, national associations, and stadium developers, to avoid developing projects out of scope and/or with the risk of not meeting their requirements and expectations. One of UEFA's key roles is to inspect stadiums before they host matches in European competitions. The inspection is used to assess the stadium's physical conditions, health and safety set-up, and security facilities and measures that are normally put in place to ensure that a match takes place smoothly (UEFA, 2023).

The development of stadia

Until the mid-1990s, the role and impact of sports stadia in shaping urban policy and sustainable design had received little attention from researchers. As tragedy struck in a number of older soccer venues and the lure of hosting major, international, sporting events captured the attention of cities and countries around the world, interest in gaining a better understanding of stadia, their place within an urban setting, their design and management, and their economic impact flourished (Bale, 1993a; Bale, 1993b; Froomberg, 1993; Sheard, 1995; Bale & Moen, 1995; Stevens & Wootton, 1997; Sayer, 2015). From this analysis, the idea that the potential of a stadium is determined by the interaction of market and socio-economic variables was clearly consistent with the five types of stadia identified in the USA by Stevens (1994) – from the earliest, classic ballpark, to the modernist super stadium, the neo-classical ballpark, the regenerated stadium, and then the Millennium Stadium.

The evolution of stadia development has been further identified by Campos (2005) and by Paramic-Salcines (2008), whose work echoed that of John & Sheard

(2000; Sheard, 2004) who plotted a number of phases in the evolution of football stadia – from the humble football ground to the post-modern stadium.

In 2004, Sheard identified four generations of stadia. He proposed that the emerging sports venues of the late nineteenth century constituted the first generation of stadia – what Leeworthy (2012) called "the 1880's wave of development and transformed sport, especially soccer, into one of the major commercial industries in the world." As football became more regulated, modest football grounds transformed into the first stadia. In Britain, many of these stadia were built in the industrialised areas of the north and funded by the factory owners who sponsored the clubs. Many of this first-generation stadia were designed by Archibald Leitch, a Scottish engineer. Inglis (2005) explains, "In the days before international stadia design specialists Leitch's engineering company was the only one in Britain to focus on football grounds. So in demand was Archie that had you attended a League match in 1939, the year he died, there was a one in three chance of sitting in a stand or standing on a terrace designed by Leitch. Starting at Kilmarnock in 1899 he was involved with no fewer than 48 grounds. Ten were new builds, including Ibrox Park for Archie's beloved Rangers (1902), Old Trafford for Manchester United (1910) and Highbury Stadium for Arsenal (1913)."

For Sheard, these stadia that began to supersede them in the post-war period were to be regarded as the second generation. The immediate post-war years of 1948–1949 in Britain saw attendance at league soccer grounds reach 41 million, and crowds of over 80,000 were not uncommon in the First Divisions in England and Scotland, yet between 1945 and 1987 only three new stadia had been built in Great Britain.

By the time we get to the Millennium in 2000, many of the structures that were evident were considered by Sheard (2004) and KPMG (2013) as the third generation. Their design and construction were highly influenced by the recommendations of the Popplewell Report (Home Office, 1986) and the Taylor Report on the Hillsborough disaster of 1989 (Home Office, 1990a) in respect of safety and security. These inquiries followed an unprecedented series of disasters at stadia between 1971 and 1989 in Scotland (Ibrox), England (Bradford and Sheffield), and Belgium (Heysel) in which over 300 people died (Goldblatt, 2013) – many of which were predicted by Hopcroft (1968).

This resulted in the publication of various advisory guidelines for safety at sports grounds and other design considerations in Great Britain and elsewhere (see, e.g. FIFA, 1991; IASLF, 1993; Home Office, 1990b; and Panstadia International, 1992). In the early 1990s, the government of Great Britain established the Football Licensing Authority and the Football Stadia Advisory Design Council together with the Stadia and Arena Management Unit that was established to specifically develop appropriate training programmes for the new generation of stadia professionals required to operate the stadia under these new guidelines.

For Inglis (1996), 1990–1996 was the most intense period of ground development in Britain since the 1890s. He says, "the 15th April 1989 at Hillsborough (in Sheffield, England), was the absolute turning point far more than any other tragedy. Almost all assumptions have been blown away: about where grounds are sited, and in what context; about how grounds are designed, planned, funded and constructed; about how they are monitored, managed and maintained, and who should operate them, own them and even use them." Inglis was dogmatic. He saw these changes as a much-needed revolution in 1996 – a revolution that had only just begun when football clubs were only beginning to be touched by the ramifications of this movement.

The new safety regulations, together with increasing commercial pressures to generate greater levels of revenue from within stadia through increased capacity, the need for ever more commercial outlets, and a desire for a new design aesthetic, began to attract international architectural companies to the world of stadia, including, for example, Herzog and de Meuron, Norman Foster & Partners, and Zaha Hadid. The demand for new stadia was given further momentum with the clamour amongst countries and cities to compete to host global soccer events (such as the FIFA World Cup, the Africa Cup of Nations, the UEFA Euro, and the UEFA Champions League). This resulted in the emergence of specialist stadia design and architectural companies, notably HOK, Populous, and gmp Architekten (Re-thinking the Future, 2023).

In 2004, Sheard (2004) had already begun to predict the developments that would shape the fourth generation of stadia to be built by 2020. These would include the new digital technologies to enhance guest experience and help generate secondary sources of revenue – from parking apps such as JustPark and Stadium Park, or VenueNext giving information about in-stadia facilities, ExApps and Pogoseat allowing fans to upgrade their seats, and online playbacks on mobile phones and 360-degree LED video screens in the stadium. There will be an ever-greater emphasis on food and drink hospitality areas, on-site and in-built hotels, extensive merchandising, heritage, and immersive visitor attractions delivering 365-days-a-year operation driven by multiple uses and higher standards of playing surfaces (both permanent and mechanically moveable).

If these developments characterised the fourth generation of stadia, then it will be sustainability that defines the fifth generation of soccer and multiple-use stadia. Sustainability in stadia will be at the heart of the new revolution determining the design and operation of stadia in line with the UN's Sustainable Development Goals (KPMG, 2013). However, the sustainable soccer stadium has been a long time coming. For example, in *Stadia: A Design and Development Guide* (John & Sheard, 2000), there is not one reference to "sustainability," and even in Sheard's seminal work on sports architecture of 2004, there is only one mention of "sustainability" (on page 63) which states, "Thus design in the broadest sense must seek to balance, integrate and reconcile all the elements of enjoyment, some of which are listed below: people, architecture, art and culture, location and sustainability being the conserving and re-using natural resources, low maintenance, non-polluting outputs and longevity of plant material." Although recycled materials are now commonly used, in 2012 the London Olympic Stadium was regarded as "an early adopter" of this technology with 40% of the concrete used being made from recycled aggregate (John & Parker, 2020).

According to Walker (2018), "In the past decade, environmentally-friendly operations and practices have gone from marginal concern to a major consideration in sports venue management *(and design?)*. Organisations – such as the Green Sports Alliance – have been set up to offer advice. There is now an understanding that sustainable stadium equals economically efficient stadium."

Examples of best practices

This trend towards incorporating sustainable building practices is now driving innovation in the design and construction of stadia. Since the first brick was laid in 1993 followed by its opening in 1996, the 56,130-capacity Amsterdam ArenA (renamed the Johan Cruiyff ArenA) – home to football, the Dutch international soccer teams, and AFC Ajax – has championed sustainability, setting an early target to be carbon-zero by 2015, and has pioneered innovative solutions. Indeed, the arena's original promotional material defined the development as being "the heart of a new sustainable urban concept for the twenty-first century" (Stevens, 2005). For example, in 2011, there was a collaboration with the Brazilian petrochemical company Braskem to use its green plastic to make nearly 2,000 new seats for the stadium. Under the agreement, Braskem supplied the raw material to make the seats from 100% renewable ethylene, derived from sugar cane and known as the "sugar seats." In 2017, a further initiative saw the arena's sustainability team working with Amsterdam University's Faculty of Applied Sciences to recycle its old plastic seats (van Dijk, 2017).

It is the arena's vision to be a state-of-the-art stadium now and in the future, where mobility, sustainability, fan experience, safety, and security are combined into an open innovation platform where development and testing of smart stadium and smart city solutions can prosper. This is part of the stadium owner's strategy to turn the Johan Cruiyff ArenA into a landmark in the city, which was aimed to become the world's most sustainable capital.

As a result, the collaboration with organizations like KPMG and Microsoft and other parties to gather more knowledge and create a vibrant ecosystem where innovation and sustainability thrive. Indeed, the Johan Cruijff ArenA believes that future smart cities will switch their focus from the optimization of individual systems to the creation of effective network systems. Linked systems based on advanced IT solutions are essential for the establishment of such smart networks.

In 2015, the ArenA began with the development of a digital marketplace as an innovative learning and development environment, open to all. They wanted to create an environment where quick advancements and smart applications can be developed, tested, and brought to life. This platform supports the development of a smart city prototype and is the foundation for a data-driven ecosystem in Amsterdam. By collaborating with KPMG, Microsoft, and other organizations, Johan Cruijff ArenA has been able to transform into a more data-driven organization. KPMG supports innovation in and around the stadium, as well as the development of the ArenA's strategy and governance.

The ArenA leverages trusted Microsoft technology, including the Microsoft Azure cloud and Microsoft 365, for its workstations, and all the data produced are stored in the Azure Data Lake, helping further drive the stadium's goals to become a data-driven operator and pioneer a data-driven city. This resulted in a flexible data analytics platform to organize, share, and enable data analysis. The approach leads to an open data-driven innovation ecosystem that anyone can join or contribute to. KPMG and Johan Cruijff ArenA will continue to actively search in the market for corporate, start-up, and scale-up partners for the innovation ecosystem, which will enable expansion possibilities.

Through this collaborative model, the Johan Cruijff ArenA is able to further strengthen its innovation ecosystem and develop applications that enhance the experience for stadium visitors. In addition, it will be possible to scale the living lab approach internationally, and the stadium and smart city consulting services can be expanded. The digital platform is designed to deliver a strong fan experience and build sustainability.

Stadium visitors can connect to the 55,000-seat stadium via smartphone for real-time directions that guide them directly from home to their seats. The system updates users on traffic issues, parking availability, and public transportation options. State-of-the-art sensors monitor all aspects of the stadium – from the grass health to crowd movement – in order to help deliver the best experience possible. The sensors also help put security in the right place at the right time to avoid challenges. It manages an emergency power system based on recycled electric car batteries that can provide power to both the stadium and surrounding businesses in case of a power outage. Fans can enjoy in-stadium video footage of key moments during play.

Johan Cruijff ArenA is well on its way to meeting its goals. Since the project began, it is close to reaching its goal of increasing revenue by 20% and has achieved its goal of 20% lower costs due to the efficiencies gained in serving customers better and faster. And, the stadium is able to operate with a carbon footprint of zero because of its energy supply programmes. The digital ecosystem is helping the stadium manage the complexity required to keep all of the partners working together in a trusted environment that fosters innovation. They are continuing to develop new ideas for the future to evolve and contribute to a world-leading smart city.

At the First World Sports Tourism Congress in November 2021, organized by the United Nations World Tourism Organisation and Agència Catalana de Turisme (2022), Croke Park in Dublin was highlighted as a benchmark for sustainable stadia. As home and headquarters of the Gaelic Athletics Association (GAA), an Irish icon and a venue for up to 1.5 million visitors a year hosting traditional GAA sports as well as occasional rugby and soccer games, Croke Park has championed the importance of sustainability globally for over 10 years. Proud of its long-established sustainability credentials, Croke Park was the first stadium in Ireland and Britain to secure both ISO 14001 and ISO 20121 standards. In another first, Croke Park was the very first stadium in the world to obtain certification to the latest international Environmental Standard ISO 14001:2015. "If we can do it in Croke Park, then it can be a model for the rest of the country" is how the GAA Commercial Director & Croke Park Stadium Director, Peter McKenna, puts it. "Culturally, it has become a core value of Croke Park to pursue best practice in sustainability and to encourage others to do so by following our example" (Stevens, 2023). GAA stars and supporters recently joined together for the launch of Croke Park's stadium sustainability day, which celebrated the stadium's bird and wildlife habitats. On the day, supporters were able to take part in Croke Park's first ever reusable cup scheme and sample dishes made from Croke Park honey.

Energy efficiency is an area of major focus for Croke Park. A state-of-the-art environmental improvement programme is in place covering the stadiums electricity, and a range of energy efficiency measures have been implemented that has led to a reduction of almost 75% in carbon emissions. In 2014, Croke Park achieved the landmark of full waste diversion away from landfill, and 100% of the waste produced is now recycled, reused, or recovered as solid fuel. Not only is 0% of the venue's waste sent to landfill, but there has also been a 12% reduction in the total amount of waste produced in the past five years with the introduction of many new compostable items throughout the stadium.

Croke Park is a member of the Water Stewardship Programme, which is accredited to European Water Stewardship (EWS) standards, designed to help reduce water consumption and introduce more efficient water management and monitoring across the facility and operations. One of the more engaging projects is Croke Park's biodiversity programme, which started in 2015 with a bug hotel and by 2018 had expanded to include the GAA's new turf farm and the placement of special bird nesting boxes throughout the stadium. The farm is used to grow herbs and vegetables for use in the stadium and manage beehives producing Croke Park honey. Croke Park was the founding partner for the Dublin North Central Garda Youth Awards. It is still going strong in 2023. Croke Park's relationship with its nearest neighbours is also an important focus for the stadium, with local community projects in place to lessen their impact of events on the community and help the area thrive. These include a community fund project that has allocated over €1 million in support to over 200 different local community groups, an event day community team made up entirely of residents, and a calendar of special projects and activities for neighbours to get involved in.

Technology is key to sustainable stadium future developments. This will involve innovative use of materials. For example, at the €80 million, 25,100-capacity Stade Océane in Le Havre, opened in 2012, the KSS Design Group with the use of ethylene tetrafluoroethylene (ETFE) allowed the venue to become Europe's first carbon-positive stadium. Used as cladding, ETFE has a carbon footprint that is much lower than comparable systems and weighs just 1%–3% of traditional materials. Wood is also making a comeback. In Italy, Bear Stadiums architectural practice is collaborating with the Italian timber manufacturer Rubner Holzbau to deliver modular, green, stadia capable of being assembled within eight months. The modular approach allows the stadium to expand incrementally as the club grows from modest beginnings of say 5,000 capacity to 20,000, thus making them economically sustainable as well (Holzbau, 2022; Holzbau, 2022a). According to Peter Rubner (2022b), "Our ecological stadiums, which are built by using the most sustainable construction material ever, score high due to their extremely low environmental impact. In addition, these installations contribute to reduce greenhouse emissions and energy consumption and can be installed – thanks to the lightweight timber construction – even in those areas subject to high seismic risks. Many football associations, which have committed themselves to implement the FIFA Sustainability Strategy and to support the Climate Neutral Now Initiative, have already expressed their interest."

Sustainable actions are also being taken at a more practical level. For example, over the weekend of 3–5 February 2023, the Professional Footballers' Association (PFA) announced its first-ever "Sustainability Champion" (PFA, 2023), and across the UK, more than 120 Premier League, English Football League (EFL), and National League clubs took part in the Green Football Weekend (2023) designed to unleash the power of football to tackle climate change. The initiative was launched in early January 2023, and millions of fans were encouraged to take climate-friendly actions that bring "green goals" for their team and were competing for the Green Football Cup. The idea is to start changing basic habits around football so that this conditions a wider cultural transformation. As part of the tournament, fans were able to go on the greenfootballweekend.com website and register "green goals" for their club through climate-friendly actions such as eating a vegetarian meal or turning a thermostat down 1 degree. Green Football Weekend will then see clubs join fans in implementing changes by making their fixtures "greener games," with some teams wearing green armbands to show their support.

Green Football Weekend is set to become an annual event and is backed by more than 30 major supporters, including the Football Association, the EFL, the Women's Super League, the National Trust, the Royal Society for the Protection of Birds, and the Church of England. It is estimated that if the UK's 36 million fans adopted three climate-friendly habits across the course of the 20-day tournament – reducing their shower time to four minutes, having two meat-free days per week, and turning the thermostat down by 1 degree – fans could save more than 700 million kg CO_2e, the equivalent of planting 11.78 million trees. Examples of how some clubs responded to the challenge include the following: Bristol City's first team players visited local primary schools to take part in an environment-focused assembly led by the club's community team; Middlesbrough conducted a boot and clothing collection at the match against Blackpool on Saturday, 4 February, with the MFC Foundation recycling old boots and distributing to communities, and clothing will be recycled via local charities; Ipswich Town launched their partnership with the Ipswich *Clear Air Now*" campaign; Leicester City created a range of new nature habitats around their stadium; Blackburn Rovers promoted sustainable options at matchday catering outlets and working with Planet League on fan and community "green goals" engagement; and Crewe Alexandra will reveal plans on Saturday to install more than 3,000 solar panels above parking spaces on the existing car park at Gresty Road, which has the potential to make the club carbon-negative.

For Jaime Manca di Villahermosa of Bear Stadiums, the major growth in demand for a sustainable stadium will come from smaller soccer clubs that need stadia with capacities of 5,000–20,000. Di Villahermosa states that 80% of all future global demand for sustainable stadia will come from this source (Walker, op cit). Investment in new stadia continues at a pace. In January 2022, CNBC reported that in the US several sports teams are planning to invest $10 billion (€9.2 billion) between 2022 and 2030. This includes the Inter Miami CF $1 billion soccer campus, the GEODIS Stadium in Nashville, Sacramento Republic's Railyards project, and the Kansas City Current Stadium.

In 2016, Red Bull put down a challenge for future sustainable stadium development by listing, as their benchmark, eight stadia designed to save the world (Red Bull, 2016), including the following:

- The Morro da Mineira Stadium, Rio de Janeiro, Brazil
- The New Lawn for Forest Green Rovers, Stroud, England
- Signal Iduna Park, Dortmund, Germany
- Princes Park, Dartford, Kent, England
- MetLife Stadium, New York, USA
- Amsterdam ArenA, Amsterdam, the Netherlands.

Conclusion

In 2021, a conversation by the author with Neil MacOmish, board director of the international design and architectural practice Scott Brownrigg, resulted in the production of two short essays on sustainability in sports (MacOmish & Taylor, 2020; MacOmish & Stevens, 2020). In part, the essay by MacOmish and Taylor was replicated in the conference proceedings of the First WSTC (MacOmish, 2022). MacOmish wrote that it must be entirely clear then that any project at the conception stage and through to completion and legacy must be considered and measured against the UN's 17 Sustainable Development Goals. Equally, it must also be recognized that there will be tensions – forces in opposition – that will make such judgements difficult. There will be clear and obvious benefits to local and national economies that may have adverse consequences on local communities. The assessment of global resources in development and in use must be rigorous. But these are not circles that cannot be squared; what is required is a new process that includes professionals, community engagement, inclusive and diverse contribution, local and national governments, and institutions. It requires a broader definition of legacy – one that might be best described by the late Professor Charles Jencks as "multivalent" (e.g. not just a sports stadium used 10 times a year, but a community hub, a college, and an educational institute, as a place for start-up companies, a museum, or a cultural facility, one that responds to a philosophy of "long-life, loose fit").

A fundamental question emerges as a result of the FIFA World Cup staged in Qatar in 2022 and plans for the 2026 FIFA World Cup being jointly hosted by the US, Canada, and Mexico that gets to the heart of the sustainable stadium discussions. Does it make sense that one global sports event promotes its response to sustainability by stating that all venues are within 40 miles of one another but requires vast open-air facilities to be mechanically cooled and ventilated, whilst another states that by reusing existing facilities it is green but promotes hundreds of aeroplane journeys to those venues that are thousands of miles apart? Neither seems to be appropriate.

Global soccer events and their stadium infrastructure can be regenerative, fundamentally change countries, regions, and communities but only if long-term and sustainable criteria are applied. The COVID-19 pandemic has demonstrated that we still need international connectivity, and travel not only broadens the mind but also facilitates economic benefits that can create positive change – but can also promote a greater understanding of cultural diversity, inclusiveness, and tolerance. Technological advancement will help us in our targets for a sustainable environment, but a realigned attitude must be the principal driver. If the global pandemic has taught us anything, it is that when we are faced with an immediate threat, we can, through collaboration and collective effort, respond quickly and effectively (whilst recognizing that there have been aspects of the response that have fallen short). Necessity, it is said, is the mother of invention. We need to be inventive on a community, cultural, and global scale now to deal with the unquestionable threat that lies before us.

This is the challenge facing the owners, investors, designers, and operators of soccer stadia. For some countries and clubs, these challenges are way beyond their immediate concerns when basic human survival is their paramount concern be it in Ukraine, Iran, Turkey, and elsewhere (O'Connor, 2020).

For the rest of us, there is still a long way to go. Given the popularity of professional sports, especially soccer, across cultures, the adoption of pro-environmental initiatives – especially by leading teams and brands – has the potential to inspire positive social change. The sustainable stadium is perhaps one of the most visible and tangible features to demonstrate this commitment. However, according to Kellison et al. (2015), disappointment reigns as, at the time of writing in 2015, less than 40% of all new stadia being built in North America had strong sustainable credentials.

Starting from the earliest examples of stadia from the Roman and Greek periods, stadia have always been significant urban elements, often built with an extravagant flair and an eye to the spectacular (Ertan & Özfiliz, 2015), and are generally symbols of a community and its culture. As a result, the design and management of stadia evolves over time to reflect cultural and societal change, especially today in relation to the effective, efficient, and sustainable use of resources. The challenge is more than one of the architectural perspectives. It is about the total implementation of a philosophy of applied sustainability that embraces the physical, environmental, social, and financial aspects of creating a sustainable soccer stadium.

Concerns about how this can be achieved date back over 30 years (AIA/UIA, 1993), and, as we have seen, calls for a sustainable approach have been repeated throughout the intervening years (Henry, 2000; Ertan & Özfiliz, 2015). The International Olympic Committee (IOC, 2021) believes that sports present broad opportunities to promote environmental awareness, capacity building, and far-reaching actions for environmental, social, and economic development across society. It can also be a means to achieving peace and reconciliation as a fundamental prerequisite for sustainability principles to be shared and applied. Nowhere is this edict better exhibited than in the brief for the development of the new Casement Park Stadium by the GAA in west Belfast in readiness for hosting the 2028 UEFA European Championships and the public engagement work currently being undertaken to shape the reality of the vision (Jenkins & Stevens, 2022).

An important step forward in measuring the overall sustainability of football clubs and their stadia has now been developed in the UK. In February 2023, Fair

Game UK introduced the Football Sustainability Index (2023). The index is, according to Fair Game's CEO, Niall Couper, "a remarkably simple concept. Football is broken. Owners are playing fast and loose with clubs and at risk is those clubs' very existence and with it decades of histories, traditions, and community projects. That culture needs to change. The Index highlights clubs that live by values that both benefit football and help shape a better future and culture for the game we all love."

In order to develop the index, Fair Game has identified four key criteria of sustainability (all of which were prominent in the recent UK government's Fan-led Review into the governance of football). The Football Sustainability Index is constructed as the weighted score of four sub-indicators of the index. The four sub-indicators and their weights are as follows:

- Financial solvency (40%): This is calculated as the weighted sum of

  - current assets/liabilities (30%)
  - short-term loans measure (25%)
  - loans repayable within one year as % of revenue (25%)
  - wages as % of revenue (20%). The data come from official accounts filed with UK Companies House.

- Governance (30%): This is calculated as the weighted sum of

  - clear governance and CSR compliance (provided by Responsiball – www.responsiball.org) with 50% weight
  - environmental measure (taken from Sports Positive Leagues' website – www.sportspositiveleagues.com) with 45% weight
  - Living Wage Employer Accreditation with 5% weight

- Fan engagement (20%): This is calculated as the weighted sum of

  - fan with 90% weight
  - percent of stadium filled on league matchdays with 10% weight

- Equality standards (10%): This is calculated by

  - ratio of women on the club board (50%)
  - recruitment ratio of women and BAME to leadership roles (provided by the Football Leadership Diversity Code – www.premierleague.com) with 50% weight. Where there were no data from the Football Leadership Diversity Code, a zero score was applied. This is also the case where it has been unclear at this stage to determine exactly how many women are among the officers listed on UK Companies House.

The index scores every professional club in the Premier League and the Championship on these four criteria, creating a clear and transparent status report. In the future, this analysis will be expanded to include all clubs down to National League North and South and the top two tiers of the Women's Game in England and Wales. The 2023 index reveals that the top five clubs in the English Premier League were Liverpool (score 70.09), Southampton (69.75), Arsenal (69.74), Tottenham

Hotspur (68.89), and Manchester United (66.50). The three Premier League Clubs achieving the lowest scores were Newcastle United (38.79), AFC Bournemouth (32.29), and Nottingham Forest (22.69).

Couper (2023) states that

> The Index brings together existing industry-leading indices and academic expertise to create a robust scoring matrix. The integrity of the Index is built on constant improvement. The data is peer reviewed by a team of experts from a range of universities. As part of the process of continuous improvement, each year the experts and Fair Game's Advisory Council will refine the methodology to improve the Index's usefulness and usability. Ultimately the aim is that the Index will be run by a new independent regulator for football, creating a new benchmark for the health of football.

This is clearly a step in the right direction. It is a model that can be further refined and more widely applied.

References

Agència Catalana de Turisme. (2022). WSTC21: 1er Congrés Mundial de Turisme Esportiu. Spain: Barcelona.

American Institute of Architects. (1993). *A call for sustainable community solutions*. Washington: American Institute of Architects.

Bale, J. (1993a). *Sports, space, and the city*. London: E. & F. N. Spon.

Bale, J. (1993b). The spatial development of the modern stadium. *International Review of Sociology of Sport*, 28, 121–133.

Bale, J., & Moen, O. (1995). *The stadium and the city*. Keele: Keele University.

Bellos, A. (2002). *Futebol: The Brazilian way of life*. London: Bloomsbury.

Birley, D. (1993). *Sport and the making of Britain*. Manchester: Manchester University Press.

Budka, P., & Jacano, D. (2013, October). Football fan communities and identity construction. *Paper at Kick It – The Anthropology of European Football*.

Buford, B. (1992). *Among the thugs: The experience, and the seduction, of crowd violence*. London: Secker & Warburg.

Campos, C. (2005). La Revuelta de los Estadios – the Spanish Stadia Revolt. *Sport Business Review*, 19, 6–13.

Col·legi d'Architectes de Catalunya. (2018). *El Camp Nou: 50 anos de latidos azulgrana*. Barcelona, Spain: Barcelona.

Couper, N. (2023). *The football sustainability index 2023. Analysis*. London: Fair Game UK

Dezeen. (2023, February 6). Retrieved from www.dezeen.com

Ertan, S., & Özfiliz, S. (2015). *Stadium construction and sustainability: The review of mega-event stadiums (1990–2012)*. Turkey: Faculty of Architecture, Middle East Technical University.

Fédération Internationale de Football Association, Union of European Football Associations. (1991). *Technical recommendations for the construction of new stadia*. Zurich: FIFA.

Filger, S. (1981). *Sport and play in American life*. Philadelphia: Saunders Press.

Fishwick, N. (1989). *English football and society*. Manchester: Manchester University Press.

Foer, F. (2010). *How soccer explains the world: The unlikely theory of globalisation*. New York: Harper.

Football Ground Guide. (2023, February 6). Retrieved from www.footballgroundguide.com

Froomberg, C. J. (1993). *Developments on the field of play*. Summer. London: Panstadia.

Goldblatt, D. (2013). *The ball is round: The global history of soccer*. New York: Riverhead Books.

Green Football Weekend. (2023, February 5). Retrieved from www.greenfootballweekend.com

Henry, P. (2000). *Design and sustainability and stadium Australia*. Sydney: UIA Sports & Leisure Group Sydney Meeting. 28 June.

Hinch, T., & Higham, J. (2011). *Sport tourism development* (2nd ed.). Bristol: Channel View Publications.

Holt, R. (1989). *Sport and the British*. London: Clarendon Press.

Holzbau, R. (2021, May). *Brief case history – modular wood systems*. Italy, Kiens: Holzbau Rubner.

Holzbau, R. (2022a). *Stadium construction in the 21st century*. Italy, Kiens: Holzbau Rubner.

Holzbau, R. (2023, April 19). *Personal correspondence to author*. Italy, Kiens.

Home Office. (1986). *Committee of inquiry into crowd safety and control at sports grounds. Chair. Mr. Justice Popplewell, final report*. London: Cmnd. 9710, HMSO.

Home Office. (1990a). *The Hillsborough Stadium disaster, 15 April 1989. Inquiry by Rt. Hon. Lord Justice Taylor, final report*. London: HMSO.

Home Office. (1990b). *Guide to safety at sports grounds*. London: HMSO.

Hopcroft, A. (1986). *The football man*. London: Aurum Press.

Inglis, S. (1993). *New directions in stadium design*. London: The Building Centre.

Inglis, S. (1996). *Football grounds of Britain*. London, UK: Collins Willow.

Inglis, S. (2000). *Sightlines: A stadium Odyssey*. London, UK: Yellow Jersey Press.

Inglis, S. (2005). *Engineering Archie*. Played in Britain/Historic England.

International Association for Sports and Leisure Facilities (IASLF). (1993). *Planning principles for sports grounds/stadia*. Koln: IAKS.

International Olympic Committee. (2021). *Sustainability through sport: Implementing the Olympic Movement's Agenda 21*. Lausanne: IOC.

Jenkins, J., & Stevens, T. (2022, August 7). *The future of community-focused sports stadia*. Paper presented to Feile an Phobhail. Belfast.

John, G., & Parker, D. (2020). *Olympic stadia: Theatres of dreams*. London: Routledge.

John, G., & Sheard, R. (2000). *Stadia: A design and development guide*. London: Architectural Press.

Kellison, T., Trendafilova, S., & McCullough, B. (2015). *Considering the social impacts of sustainable stadia design*. Retrieved on 27th December, 2022 from www.researchgate.net/publication/281378264

KPMG. (2013). *A blueprint for successful stadium development*. London: KPMG.

Leeworthy, D. (2012). *Fields of play: The sporting heritage of Wales*. Aberystwyth: Royal Commission on the Ancient and Historical Monuments of Wales.

MacOmish, N. (2022). *Climate, community, culture: Tensions and oscillations*. UNWTO.

MacOmish, N., & Stevens, T. (2020). The Future of Global Tourism. In *Intelligent architecture* (Issue 11). London: Scott Brownrigg.

MacOmish, N., & Taylor, H. (2020). Pure research: Returning to stadia. In *Intelligent architecture* (Issue 11). London: Scott Brownrigg.

O'Connor, R. (2020). *Blood, and circuses: A football journey through Europe's rebel republics*. London: Biteback Publishing.

Panstadia International. (1993). *Leading the field: A worldwide guide to stadium new-build and management*. Harrow: Panstadia International.

Parmic-Salcines, J. (2023, January 28). *From modern to post-modern: The development of football stadium*. Retrieved from www.researchgate.net

Professional Footballers' Association (PFA). (2023, February 2). Retrieved from www.thepfa.org

Re-Thinking the Future. (2023, February 7). *Architects reshaping the world of stadia*. Retrieved from www.re-thinkingthefuture.com

Red Bull. (2016). *Eight stadia to save the world*. Retrieved from www.redbull.com

Ritchie, B. W., & Adair, D. (2004). Sport tourism: Interrelationships, impacts and issues. In *Aspects of tourism*. Bristol: Channel View Books.

Sandry, A. (2016). *Opening address*. Gwlad, Gwlad: Football and Identity Conference, School of Management, Swansea University.

Sayer, J. (2015). *Role of football stadia in Britain's built environment*. Nottingham: University of Nottingham. May.

Sheard, R. (1995). *A stadium for the nineties*. London: Panstadia, Spring.

Sheard, R. (2004). *Sports architecture*. London: SPON Press.

Stevens, T. (1994). Stadia and Arenas: The sleeping giants of tourism. In A. Seaton, et al (Eds.), *Tourism: The state of the art*. London: Wiley & Sons.

Stevens, T. (2005). Sport and urban tourism destinations: The evolving sport, tourism, and leisure functions of the modern stadium. In J. Higham (Ed.), *Sport tourism destinations*. Oxford. Elsevier Books.

Stevens, T. (2023, January 17). *Personal interview with Peter McKenna*. Dublin.

Stevens, T., & Wootton, G. (1997). *Sports stadia and Arena: Realising their full potential* (Vol. 22. Issue. 2). Lucknow. India: Tourism and Recreation Research.

Strutt, J. (1801). *Glig-gamena angel deod, or, the sports and pastimes of the people of England*. London New edition 1903, re-printed Bath 1969.

UEFA. (2023, February 2). Retrieved from www.uefa.com

UEFA. (2018). *Stadium infrastructure regulations edition 2018*. Zurich: UEFA.

UNWTO and Agèncie Catalana de Turisme. (2022). *WSTC21: 1er Congrés Mundial de Turisme Esportiu*. Spain. Barcelona.

Van Dijk, L., et al. (2017, October). Urban technology. *The Second Life of a Stadium Seat: The Amsterdam ArenA Case Study*. Amsterdam University Applied Sciences.

Walker, T. (2018). The future of stadia. In *Sports management* (Issue 1). Hitchin, Hertfordshire UK: Leisure Media.

Weed, M., & Bull, C. (2004). *Sports tourism: Participants, policy, and providers*. London: Elsevier.

Other useful sources

Retrieved from www.footballgroundguide.com
Retrieved from www.nssmag.com
Retrieved from www.sportsplanningguide.com

8 Sports tourism and event legacy

Building inclusive, safe, and resilient cities and communities

Dušan Borovčanin

Introduction

The concept of legacy is *"rather a dream to be pursued than a certainty to be achieved"* concluded Jean-Loup Chappelet in his review of mega sporting events and legacies 10 years ago (Chappelet, 2012). Event legacy is often discussed as a way to create safe, resilient, and inclusive cities and communities. But as Chappelet stated, indeed, more than a decade ago, the legacy concept is still far from full implementation in various examples of mega events, not to mention the other forms and sizes of the events.

While one could argue we are still not close to fully collecting benefits from the legacies of the past events, a decade later, there is significant progress in concept development of creating a legacy from sporting events, and not merely from sporting events but from events of any kind, and especially business events (Foley et al., 2013). A recent study by Preuss (2019) further developed the legacy framework, adding the value with practical application, yet again underlining the difficulties in legacy measurement. Indeed, legacy measurement proved to be one of the very complex areas of legacy. Moreover, the potential for corruption and increased scrutiny by a variety of stakeholders led to a decrease in support and trust in the benefits of hosting mega events such as Olympic Games, for instance (Preuss, 2019). These increased concerns and decreased support were not baseless. There were numerous examples of hosting mega events that were connected to missed opportunities for urban development or even increased crime rates (Campaniello, 2013).

Nevertheless, the potential and the need for making mega events more sustainable and inclusive are not losing pace with the raising concerns. On the contrary, the amount of knowledge collected and public awareness of the potential costs and benefits of hosting a mega event should even improve the development of the legacy programme – in a way it can create lasting outcomes for the community and the destination as a whole.

Sports tourism, events, and legacy

Sports tourism, events, and legacies are closely connected. Sports tourism is in itself a relatively young discipline of research, and the legacy of sports tourism and events

DOI: 10.4324/9781003384786-8

is an even more recent concept. Indeed, although people engaged in sports-related travel for centuries, the popularity of these forms of travel, which is largely connected to sports events, gained much of the popularity in the last 3–4 decades. One of the first areas of interest related to this phenomenon was its impacts on the economy. Yet, it was very clear that sports events often create opportunities for social interactions and cultural exchange as well, resulting in strengthened bonds among residents and enhanced community pride (Borovčanin & Lesjak, 2021; Getz & Page, 2019).

Although it was first mentioned in 1956 (Leopkey B, 2008), the concept of legacy, or perhaps, more accurately, legacies, also started to gain popularity during the 1990s within the sports management studies (Chappelet, 2012). It was because of the increasing concerns of the costs and benefits of hosting mega events, not only from an economic point of view but also from the social and environmental aspects. Simply put, local community started questioning the cost–benefit from hosting mega events and showed interest in the consequences on the long run.

While there is a diverse body of research related to several different types of mega events, probably the biggest attention has been pointed to the previous editions of the FIFA World Cup. These studies include a wide range of evaluations, from the economic impact (Horne & Manzenreiter, 2004; Lee & Taylor, 2005), local development (Vanwynsberghe et al., 2013), destination branding and image leveraging (Grix, 2012), and social and sustainable development (Cornelissen et al., 2011; Herbold et al., 2020; Pillay & Bass, 2008).

With numerous studies putting into question the cost–benefit of hosting mega events, social concerns started to grow and question the willingness of their governments to bid for the host of mega events. After several withdraws from the bidding process, and with a significant decreased interest in bidding for Summer Olympics, the International Olympic Committee (IOC) decided to overhaul the bidding procedure, suggesting the host cities and countries should make a referendum before making a final decision on whether to bid or not (Grohmann, 2019).

Today, thinking about the legacy of a mega event from the very start, the bidding process is essential, and it makes part of the formal procedure for many mega events. Legacy is most often referred to as the long-term or permanent outcome for the host city and community from staging the event (Thomson et al., 2013). It encapsulates the lasting impacts it leaves on the host city, encompassing the social, economic, environmental, and infrastructural dimensions. These impacts transcend the immediate economic gains associated with the event and encompass a broader spectrum of lasting transformations. For instance, the Calgary Winter Olympics in 1988 spurred the development of a robust volunteer culture that endures to this day, cultivating a sense of belonging and unity (Ritchie & Adair, 2004). The concept of legacy needs to be developed from the earliest stages of the event, in order to allow every stakeholder to properly plan and develop a clear vision of what is to be achieved (Thomson et al., 2013).

Using sports, tourism, and events to make cities more inclusive, safe, resilient, and sustainable

In recent decades, the concept of urban development has evolved beyond mere infrastructural expansion to encompass broader socio-economic and environmental considerations. Cities are now recognized as dynamic ecosystems that require strategic planning to ensure inclusivity, safety, resilience, and sustainability. The integration of sports, tourism, and events into urban development strategies has gained prominence as a multifaceted approach to achieving these goals. This chapter delves into the multifaceted benefits of utilizing sports, tourism, and events as tools for creating more inclusive, safe, resilient, and sustainable cities using the process of the creation of the bid for EXPO 2027 in Belgrade with the theme "Play for humanity: Sport and music for all." International sports and mega events can help achieve SDG 10 in four major areas.

- **Promoting Inclusivity through Sports, Tourism, and Events**

Inclusivity is the fact of including all types of people, things, or ideas and treating them all fairly and equally (Cambridge Dictionary, 2023). It involves providing equal access and opportunities to all segments of society, irrespective of their socio-economic status, gender, age, or abilities. Sports, tourism, and events can serve as powerful platforms to foster inclusivity by creating spaces where diverse communities can come together and participate. For instance, hosting international sporting events often involves infrastructure development and improvements that can benefit the entire community. The transformation of Rio de Janeiro's favelas in preparation for the 2016 Summer Olympics is a prime example of how sports-driven development can enhance inclusivity (Steinbrink, 2013).

- **Enhancing Safety and Security through Sports, Tourism, and Events**

Safety and security are paramount in urban planning and development. Incorporating sports, tourism, and events into city planning can contribute to enhancing public safety. Well-organized events and tourism activities often necessitate improved infrastructure, better lighting, and increased police presence, thereby making urban areas safer for residents and visitors.

- **Building Resilience via Sports, Tourism, and Events**

Resilient cities are better equipped to withstand and recover from various shocks and stresses, such as natural disasters and economic downturns. Sports, tourism, and events can contribute to urban resilience by diversifying local economies and supporting community development. For instance, Barcelona's successful transformation after hosting the 1992 Summer Olympics not only improved the city's infrastructure but also enhanced its ability to weather economic challenges (Hall, 1993). The influx of tourists during major events can provide a buffer against economic instability and boost local businesses.

- **Advancing Sustainability through Sports, Tourism, and Events**

Sustainability is a crucial consideration in contemporary urban development. Sports, tourism, and events can play a pivotal role in promoting sustainable practices. By investing in eco-friendly infrastructure and transportation systems, cities can reduce their carbon footprint. The adoption of sustainable practices during major events, such as waste reduction and energy-efficient operations, can inspire long-term changes in the community (Gibson, 2017). Moreover, sustainable tourism practices, such as promoting responsible travel and supporting local economies, can contribute to the overall sustainability of the city (UNWTO, 2019).

The case study of Specialized EXPO 2027 Belgrade

EXPOs have long been celebrated as forums for cross-cultural dialog, technological innovation, and economic advancement. The legacy of EXPOs extends beyond the event itself, as showcased by the profound impact of the 1851 London Crystal Palace Exhibition (Bureau International des Expositions, 2023), which not only revolutionized architecture but also epitomized the transformative potential of international exhibitions These events cultivate a legacy of increased international cooperation and technological progress, exemplified by the legacy of the 2010 Shanghai EXPO, which catalyzed sustainable urban development and technological innovation (Xu, 2013).

Both EXPOs and mega sports events often necessitate extensive infrastructural development, leaving a physical legacy that contributes to urban transformation. The Barcelona 1992 Olympics stands as a testament to the infrastructural prowess of mega events, with the city's rejuvenated waterfront and transportation systems continuing to shape its urban fabric (Hall, 1992). Similarly, the EXPO 2015 in Milan led to the creation of a new urban district and sustainable infrastructure, leaving a legacy of enhanced public spaces and improved connectivity (De Carlo et al., 2009).

The legacy of Serbia as a participant at the world fairs

> Serbia has a long history of appearing at world exhibitions and celebrates over 138 years since the first World Exhibition in Antwerp, Belgium (1851), when the Kingdom of Serbia exhibited alongside France, Germany, Austria, Canada, Great Britain, the Ottoman Empire, Portugal, Spain, the Kingdom of Romania and the United States of America. Serbia's pavilion occupied an area of 125 square meters, providing space for 302 exhibitors, and, during the exhibition, the country's representatives won a total of 157 awards.
>
> (Serbia Creates, 2023)

World fairs have afforded Serbia a unique avenue to promote its cultural heritage and artistic accomplishments. Participation in these events has allowed Serbia to present its traditional crafts, music, literature, and visual arts to an international audience, contributing to the preservation and promotion of its national identity (Adamović, 1996; Hajdu, 2015). For instance, Serbia's presence at the 1939 New York World Fair showcased its cultural richness and artistic prowess, capturing the attention of visitors and fostering cross-cultural interactions.

However, this was the first time in almost 140 years long history of participation that Serbia decided to bid for the host. The competition to host the Specialized EXPO in 2027 has been fierce. Five countries (the US, Thailand, Argentina, Spain, and Serbia) competed for the host. At the moment of writing this article, it is known that Belgrade and Serbia won the vote in the Bureau Internationales des Exposition, which probably makes this case even more relevant.

Specialized EXPO 2027 Belgrade

Belgrade is the capital city of Serbia. Formally, Serbia expressed interest to host the EXPO in 2027 in January 2022 (Bureau International des Expositions, 2023). After the call for candidates was closed, there was a deadline of 6 months to handle the entire bid dossier. There are certainly multiple factors that influenced the final decision. Singled out below are some of the leading principles that were embodied in the making of the bid from the start.

Development principles of the EXPO site and venue

The Republic of Serbia is a landlocked country in Southeast Europe. It is located at the crossroads of major road, rail, and river networks effortlessly connecting Western and South Europe toward the Eastern Hemisphere, Middle East, and Asia. The key was to create a sustainable concept that would allow the future use of the EXPO site and make a value for the city, which will benefit from its infrastructure as the foundation for future Belgrade growth.

The legacy created by the EXPO can transform a city by renewing urban areas and integrating new and innovative structures that improve a host city's vibrancy and quality of life.

PRINCIPLE 1: FUTURE USE AS THE NEW BELGRADE FAIR

The EXPO venues have been designed in a way that they can easily be converted into the new Belgrade Fair once EXPO 2027 is over. The master plan and other design details were developed together with the Belgrade Fair representatives and technical support team. This has ensured that the necessary functionalities and operational needs for smooth day-to-day operations of the future trade fair are planned and fully covered.

Established in 1937, the Belgrade Fair has a long and successful history and celebrated its 85th anniversary in 2022. Recent results show an increasing trend in business development and financial performance, with 2019 being the most successful year in its history. In line with the progress of the Serbian and regional economy, further growth of Belgrade Fair's business is expected in the forthcoming period. To meet these expectations, there is a need for an increase of the exhibition space capacity and for its technical and technological improvement.

With development of the largest and the most advanced and modern exhibition space in the region, the new Belgrade Fair will become one of the corner stones of future economic growth and development of Belgrade and Serbia. It will help position Belgrade and Serbia on the world map of trade fairs and contribute to the achievement of the goal of becoming the leading economy of the broader Balkan and Southeast Europe regions.

PRINCIPLE 2: LEGACY TO THE CITY

Today, the Belgrade Fair resides in one of the most attractive locations along the Sava River. The relocation of the present Belgrade Fair to the new location after EXPO 2027 will allow the city of Belgrade to use the land for waterfront development in the city center. This relocation is fully aligned with the city of Belgrade's strategy to relocate all the industrial content out of the city center. It will allow the city of Belgrade to reclaim approximately 2 km of the Sava River waterfront and to use it in a more appropriate manner, the one corresponding to the modern sustainable city. At the same time, this enables the combination of urban renewal with modern needs of spatial planning with integrative conservation in accordance with architectural heritage scope and value of the current Belgrade Fair complex.

PRINCIPLE 3: STRATEGICALLY POSITIONED – TRANSPORT-ORIENTED

The location is easily accessible to both international and national visitors. The proximity to the highway, airport, train, and bus stations will ensure easy access for EXPO and future fair use.

PRINCIPLE 4: SUSTAINABLE

The project is sustainable from all aspects – the future use of the site itself, easy access by the eco-friendly public transport, creating the foundation for sustainable city growth, and the design itself. The EXPO site will interact with the neighboring National Stadium. The two complementary uses will support each other with the necessary facilities such as conference and media centers, hotels, parking spaces, and other infrastructure.

The project has been designed to fit the site, using natural blue and green surroundings as the inspiration and energy source.

PRINCIPLE 5: INTERNATIONAL TOURISM AND SPORTS DESTINATION

The site will be integrated with sports, educational, hospitality, and other complementary uses, improving tourism and experiences for international, regional, and local travelers. If the site uses its strategic position, it can become a competitive international tourist and sports destination, easily hosting international events.

PRINCIPLE 6: GENERATOR FOR CITY EXTENSION

Infrastructure development at the site and around it will be a foundation for future Belgrade growth. The surrounding land could become a playground for planners and architects and their visions of extending the city along the Sava River to the site. It ensures sustainable city growth in line with the strategic goal to grow together with nature.

Design for all with "Sports and music for all"

The basic principles of the "design for all" approach will be especially respected, to ensure easy access to all the contents that will be available to the visitors of the exhibition. These are principles that primarily relate to easy access to means of transport, buildings, and appropriate facilities, which ensure the smooth flow of visitors, regardless of their category (elderly people, people with disabilities, parents with babies, etc.). In that sense, appropriate elevators and ramps will be installed for easier access, and special attention will be paid to the safety of cyclists and pedestrians, for which appropriate approaches, elevators, and bridges over roads will be built. In that sense, Belgrade will be a city that is open to everyone.

ENVIRONMENTAL PERSPECTIVE BASIC APPROACH – FULL RESPECT
FOR THE ENVIRONMENT

The location of the EXPO will be a completely newly built area, so the application of the most modern standards in environmental protection is desirable and possible and, at the same time, is one of the key commitments of the organizer. One of the guiding ideas in developing the concept of the exhibition space was the use of renewable energy sources, use and sustainable management of water resources (given the proximity of the Sava River site), respect for zero emissions, and integration of the site into the environment by implementing technologically advanced and long-term sustainable solutions.

The location of the exhibition itself and its immediate surroundings will be maximally greened by the implementation of landscaping solutions, which will further enable the so-called passive cooling, and appropriate water channels will be used for additional cooling of the space. In addition, the master plan envisages surfaces under roofs and pergolas, and appropriate fans will be used with a combination of water spray to cool visitors, having in mind the period of the year when the exhibition is planned to be held. The principle of protection and improvement of the environment is at the heart of the presented concept of the master plan for EXPO 2027. The use of modern technologies will ensure long-term preservation and improvement of the wider surrounding area and contribute to achieving the Sustainable Development Goals (SDGs).

SITE DESIGN LEVERAGING ASPECT OF THE ENVIRONMENT

Belgrade EXPO 2027 has a unique opportunity to become the world's first water and carbon-neutral event of a larger scale. An auspicious combination of the locality and timing of the EXPO, which will take place at the peak rainfall and sunlight period, means that coupled with the innovative and responsible building design and operation, the entire energy and water demand can be met by harvesting and reusing renewable resources. Because it is built on a former agricultural land next to a major river, the site will need to be provided with water attenuation systems to ensure balanced rainwater drainage flows. The organizer stated it will combine the attenuation/rainwater harvesting system with a wastewater processing plant to ensure that all potable and technical water requirements of up to 20,000 daily visitors, staff, and all hotel residents are met.

By providing a combination of technical and nature-based solutions, all rainwater, graywater, and black water will be processed and returned to the appropriate elements of the water supply system – from drinking and potable water to flushing and irrigation requirements. The water harvesting and recycling system will be integrated into the central cooling plant: Water source heat pumps will provide cooling for the exhibition halls by collecting the coolth from the water circuit and disposing waste heat into the domestic hot water generation for both the EXPO and hotel use. This circular approach will allow 20 pumping stations around the site to deliver 40 megawatts of peak cooling using a combination of river and harvested water.

The interior temperatures will be regulated by a low energy displacement ventilation system: Cool air will be delivered at a low level through furniture/benches, ensuring that only the occupied zone is kept comfortable and delivering 70% energy saving compared to a more conventional system. Plentiful natural daylight in the summer months will be carefully managed to minimize the need for artificial lighting while attenuating the brunt of the direct solar load. The EXPO will combine passive and low energy climatic systems and try to minimize the reliance on mechanical environmental control systems and ensure the equivalent annual energy consumption of the halls is below 70 kWh/m^2 of electrical energy consumption and will target the LEED Platinum rating.

All electricity for the site will be delivered by an extensive roof-integrated solar array and a combination of battery and green hydrogen energy storage: We propose to install over 6 megawatts of photovoltaic capacity generating 7 gigawatts of power during the EXPO opening days alone – that is, twice the projected electricity consumption of the EXPO, and the surplus will be delivered back to the Serbian grid.

WEATHER DISASTER PREPAREDNESS

The climate in Serbia is relatively mild, and there have been almost no extreme weather incidents in recent history. In summer, occasional thunderstorms with rain showers can occur, followed by strong winds in short windows. Therefore, no material risk exists regarding this issue. However, considering that the agricultural land will be converted into a development area, strict controls and measures concerning the water drainage systems and wind protection will be implemented to mitigate any risk from flood damages and wind disasters.

For the purpose of controlling the risk of floods, damage control caused by torrential rains, and other possible problems caused by water, appropriate maps and risk management plans will be prepared, which will take into account, in addition to the above, flood risks due to river and groundwater overflow, monitoring water levels and wind movements/hits.

INCLUSIVE PARTICIPATION AT EXPO 2027 IN BELGRADE

EXPO 2027 Belgrade is all about the power of play, sports, and music and their capacity to be the catalyst of change and help humanity become more resilient to the adversities of unpredictable times.

EXPO 2027 addresses the issues of the decline of play, pandemic of physical inactivity, sedentary behavior, screen time addiction, and rise in attention deficit hyperactivity disorder (ADHD) cases worldwide and across generations.

Studies show that people in low- and middle-income countries have the highest risk of developing chronic and metabolic diseases and developing these conditions at a younger age and suffering longer. They have shorter life expectancy than people in high-income countries due to the development of unique disease profiles. Moreover, they have limited and untimely access to remedies for such diseases and an uneven understanding of the benefits of physical activity.

Play, sports, and music primarily contribute not only to physical and mental health and well-being, good-quality education, and gender equality but also to the development of vital and transferable life skills, thus increasing the chances of employment, improving the financial and economic status, and making participants more willing to volunteer in the community.

EXPO 2027 Belgrade aims to engage the world in partnership with public and private organizations to join forces to create a resilient, sustainable, and prosperous future for the most vulnerable.

Therefore, the broad participation of developing countries is considered a critical success factor for the organizer.

The Serbian government stated it will support the participation of developing countries in EXPO 2027. The organizer has envisaged several types of assistance to provide comprehensive support, specific to each distinct group of countries characterized by similar needs, ensuring all the necessary means for smooth operations of their exhibitions.

Assistance will support the implementation of exhibits throughout all the stages of the EXPO, from the planning phase until closure and dismantling, thus ensuring the development and presentation of the theme compliant with the procedures and rules set out by the Bureau International des Expositions.

The subject measures will abide by the basic principles for assistance eligibility, allocation criteria, and the number of countries that will qualify for the aid.

Should the interest exceed the number of expected participants, the government of the Republic of Serbia will explore alternative funding sources for the additional needs through commercial channels. In addition, the organizer may further segment eligible participants (except for the least developed countries (LDC) according to the UN and re-assess limits to ensure just allocation of funds.

Conclusion

While one could argue the statement from Chappelet from the beginning of this chapter is still relevant and true (Chappelet, 2012), it could be argued as well that the world today is full of uncertainties and insecurities and not related to EXPOs only. This chapter tried to use the interesting bidding file from Belgrade from the candidature process for EXPO 2027 that resulted in success. The theme of EXPO 2027 is "Play for humanity: Sport and music for all." Serbia has been participating at the World EXPOs for almost 40 years, yet this was the first time it decided to run as a host, and surprisingly won.

The chapter outlines how the legacy program, that is, the idea to leave long-lasting positive outcomes for the city and local community, was embodied since the moment of project development and how it influenced the delegates in their decision-making process. It showcases the good practice of thinking and putting legacy in place from the early stage of the project development.

The host country assured to devote financial resources to make the EXPO as inclusive as possible, allowing 50–80 less developed countries to participate.

Furthermore, the project has been developed in line with environmental recommendations. Belgrade EXPO 2027 has a unique opportunity to become the world's first water and carbon-neutral event of a larger scale. Finally, strict controls and measures concerning the water drainage systems and wind protection will be implemented to mitigate any risk from flood damages and wind disasters.

Of course, there are numerous risks facing this project, since its full implementation is about to be seen in 2027. Nevertheless, this case was used to highlight the importance of understanding mega event legacy both conceptually and practically.

References

Adamović, M. (1996). "Retrospective" section in the Serbian Pavilion at the 1911 universal exposition in Rome: An artistic cross-section of the period. *Balcanica*, 301–314.

Borovčanin, D., & Lesjak, M. (2021). Sportski turizam (1st ed). Belgrade, Serbia: Singidunum University.

Bureau International des Expositions. (2023, September 8). *About EXPO's [WWW document]*. Retrieved from www.bie-paris.org/site/en/about-world-expos

Cambridge Dictionary. (2023, September 8). *Inclusivity [WWW document]*. Retrieved from https://dictionary.cambridge.org/dictionary/english/inclusivity

Campaniello, N. (2013). Mega events in sports and crime: Evidence from the 1990 Football World Cup. *Journal of Sports Economics*, 14, 148–170. https://doi.org/10.1177/1527002511415536

Chappelet, J.-L. (2012). Mega sporting event legacies: A multifaceted concept. *Papeles de Europa*, 0, 76–86. https://doi.org/10.5209/rev_PADE.2012.n25.41096

Cornelissen, S., Bob, U., & Swart, K. (2011). Towards redefining the concept of legacy in relation to sport mega-events: Insights from the 2010 FIFA World Cup. *Development Southern Africa*, 28, 307–318. https://doi.org/10.1080/0376835X.2011.595990

De Carlo, M., Canali, S., Pritchard, A., & Morgan, N. (2009). Moving Milan towards Expo 2015: Designing culture into a city brand. *Journal of Place Management and Development*, 2, 8–22. https://doi.org/10.1108/17538330910942762

Foley, C., Schlenker, K., Edwards, D. & Lewis-Smith, L. (2013). Determining business event legacies beyond the tourism spend: An Australian case study approach. *Event Management*, 17, 311–322. https://doi.org/10.3727/152599513X13708863378079

Getz, D., & Page, S. (2019). *Event studies: Theory, research and policy for planned events* (4th ed.) London: Routledge. https://doi.org/10.4324/9780429023002

Grix, J. (2012). 'Image' leveraging and sports mega-events: Germany and the 2006 FIFA World Cup. *Journal of Sport & Tourism*, 17, 289–312. https://doi.org/10.1080/1477508 5.2012.760934

Grohmann, K. (2019). *IOC overhauls bidding process for games to stop dropouts.* Reuters.

Hajdu, A. (2015). The pavilions of Greece, Serbia, Romania and Bulgaria at the 1900 Exposition Universelle in Paris. Balkan Heritages. Routledge.

Hall, C. M. (1993). *Hallmark tourist events: Impacts, management and planning* (1st ed.). Chichester: Belhaven Press.

Herbold, V., Thees, H. & Philipp, J. (2020). The host community and its role in sports tourism – exploring an emerging research field. *Sustainability*, 12, 10488. https://doi.org/10.3390/su122410488

Horne, J. D. & Manzenreiter, W. (2004). Accounting for mega-events: Forecast and actual impacts of the 2002 Football World Cup finals on the host countries Japan/Korea. *International Review for the Sociology of Sport*, 39, 187–203. https://doi.org/10.1177/1012690204043462

Lee, C.-K. & Taylor, T. (2005). Critical reflections on the economic impact assessment of a mega-event: The case of 2002 FIFA World Cup. *Tourism Management*, 26, 595–603. https://doi.org/10.1016/j.tourman.2004.03.002

Pillay, U. & Bass, O. (2008). Mega-events as a response to poverty reduction: The 2010 FIFA World Cup and its urban development implications. *Urban Forum*, 19, 329–346. https://doi.org/10.1007/s12132-008-9034-9

Preuss, H. (2019). Event legacy framework and measurement. *International Journal of Sport Policy and Politics*, 11, 103–118. https://doi.org/10.1080/19406940.2018.1490336

Ritchie, B. W., & Adair, D. (2004). *Sport tourism: Interrelationships, impacts and issues.* Bristol, UK: Channel View Publications.

Serbia Creates. (2023, September 8). EXPO Serbia | Serbia at world exhibitions [WWW document]. *Expo Serbia.* Retrieved from https://exposerbia.rs/en/expo-history-serbia.php

Steinbrink, M. (2013). Festifavelisation: Mega-events, slums and strategic city-staging – the example of Rio de Janeiro. *DIE ERDE – Journal of the Geographical Society of Berlin*, 144, 129–145. https://doi.org/10.12854/erde-144-10

Thomson, A., Schlenker, K. & Schulenkorf, N. (2013). Conceptualizing sport event legacy. *Event Management*, 17, 111–122. https://doi.org/10.3727/152599513X13668224082260

Vanwynsberghe, R., Surborg, B. & Wyly, E. (2013). When the games come to town: Neoliberalism, mega-events and social inclusion in the Vancouver 2010 Winter Olympic Games. *International Journal of Urban and Regional Research*, 37, 2074–2093. https://doi.org/10.1111/j.1468-2427.2012.01105.x

9 Young adults' views of sustainable ski destination development

Insights from cluster analysis

Kir Kuščer, Sašo Sever, and Daša Farčnik

Introduction

Sustainable tourism has been the focus of research for about three decades (see, e.g., Buckley (2012) for the review for the first two decades and Wondirad (2019) and Khanra et al. (2021) for a more recent systematic literature review). Tourist destinations have, therefore, focused on their sustainable development. This involves a balanced approach that meets the needs of current tourists and host communities while preserving and enhancing the cultural, environmental, and socioeconomic values of the destination for future generations.

Ski destinations – destinations that attract visitors primarily for skiing and related activities (Bausch & Gartner, 2020) – also face particular challenges when it comes to sustainability, given their reliance on natural snow, high energy demands, and potential impacts on local ecosystems. Climate change has actually increased the challenges for ski destinations, especially in terms of providing enough snow (Steiger, 2012; Bausch & Gartner, 2020), making the ski destinations commit to Sustainable Development Goals (SDGs), especially SDG 13 – Climate action (UN, 2019). Ski resorts and destinations are taking steps to minimize their carbon footprint by focusing on snow management, energy efficiency, and the use of renewable energy. They also strive to preserve local ecosystems and engage responsibly with stakeholders through waste management, conservation measures, and community engagement. Adopting sustainability practices can help ski resorts reduce their environmental footprint, improve guest experience, and ensure their long-term profitability.

However, supply-side sustainability practices are not enough, and sustainable travel patterns should also be promoted. This is also in line with the SDGs, especially SDG 12 – Responsible consumption and production, which especially holds for ski destinations which, due to their dependence on nature, need to develop by considering the natural environment. Although tourists increasingly recognize the importance of preserving the environment and supporting local communities for the long-term viability of tourism destinations, they have different attitudes toward sustainable travel (see, e.g., Crouch et al., 2005; Juvan and Dolnicar, 2014). Thus, several studies have found that tourists can be segmented in terms of their attitudes toward sustainable travel or their sustainable travel behavior (see, e.g., Dolnicar

DOI: 10.4324/9781003384786-9

and Leisch, 2008; Kastenholz et al., 2018). Buffa (2015) also notes that young tourists can also be segmented in terms of their attitudes toward sustainable travel or their sustainable travel behavior. In general, young people have greater awareness of sustainability (UNWTO, 2016), which could also be due to the fact that they are yet to experience the impact of potential sustainability efforts (Hill & Lee, 2012). It is important to identify different segments in order to actively target them to reduce the footprint of their travel (Dolnicar, 2006).

The sustainability challenges of ski destinations and the different attitudes toward sustainable travel motivated this chapter, in particular, the different attitudes of young tourists, who represent a significant population in the global travel industry, toward sustainable travel. Therefore, the aim of this study was to investigate young adults' attitudes toward sustainability when choosing a ski destination and their views on sustainable development of ski destinations. More specifically, we were interested in identifying the segments of young adults (aged 18–34 years) in terms of different attitudes toward the sustainable development of a destination (20 elements) and their influence on the choice of a ski destination (16 elements). A total of 267 responses were collected from young adults from EU member states, with the sample being representative in terms of EU regions. First, to examine the components of the elements that influence respondents' decision to choose a ski destination, a principal component analysis (PCA) was conducted. Second, a cluster analysis was conducted to examine the number of clusters and their attitudes toward sustainable development of a ski destination, as well as the importance of sustainability elements in choosing a ski destination.

The remainder of the chapter is organized as follows. First, we present the literature review that served as the basis for the research questions and the development of the questionnaire. Second, we present the research questions and the data and methodology, followed by the results, and finally a discussion, conclusions, and implications for practice.

Literature review

Sustainable development of tourist destinations has been the center of attention of numerous studies in recent years. Many researchers stress the need for alternative strategies, especially from the supply side, and a shift toward sustainable business models (e.g., Luthe & Schläpfer, 2011; Verbeek et al., 2011). From the demand side, there is a need for more sustainable travel patterns (Dolnicar, 2006; Fredman & Margaryan, 2021). Sustainable travel is a precondition and requirement for more responsible consumption and production – which is one of the SDGs (Seeler et al., 2021).

Tourism market is nowadays impacted by exogenous forces like climate change and demographic change (see, e.g., Steiger, 2012; Škare et al., 2021). This is particularly evident in ski destinations, since climate change has a strong influence on this weather-affected industry, especially during the winter season (Bank & Wiesner, 2011; Unbehaun et al., 2008; Steiger, 2012). Climate change is causing glaciers to melt and snow seasons to shorten. This is having a negative impact on winter tourism activities, such as skiing and snowboarding, especially due to the lack of snow reliability (Steiger et al., 2022; Bausch & Gartner, 2020). The survival of ski destinations is thus dependent on creating competitive advantages by product and consumer diversification while at the same time using sustainable development principles as the only acceptable concept (Flagestad & Hope, 2001; Škorić, 2008). Therefore, ski destinations are becoming more and more committed to contribute to SDGs (UN, 2019). Dragović and Pašić (2020) list several sustainable development practices of ski destinations directed to contribute to the sustainability of ski destinations. Knowles (2019) finds that although the North American ski industry leaders rate the climate change consequences differently, they believe their visitors value sustainability. Furthermore, Nordin et al. (2019) note that in Åre, Sweden, and Whistler, Canada, sustainability is taking a more central role in destination governance.

On the other hand, there are very few studies (e.g., Gilg et al., 2005; Miller et al., 2010; Wehrli et al., 2012) concerning visitors' understanding of sustainable development of ski destinations and the importance of different sustainability elements from their perspective. Wehrli et al. (2012) conclude that sustainability-aware tourists are an interesting market segment, for whom sustainability is among the most important elements when choosing a destination.

Even scarcer is research targeted at young adults' attitudes toward sustainability. Young travelers are often referred to as "youth tourists," are a significant demographic population in the global travel industry, and are recognized as one of the fastest growing segments of international tourism (UNWTO, 2011; UNWTO, 2016) since they are expected to travel even more as their incomes grow (Nielsen, 2017). However, the importance of this market segment is not only in the size but also in the travel behavior of the young tourists as it leads to significant changes in the tourism market (OECD, 2018) and shows the market of the future (Vukic et al., 2015). Hill and Lee (2012) further note that young adults will live on to see the implications of sustainability efforts, so their views on sustainability issues are especially important.

In addition, young people have greater awareness for sustainability (UNWTO, 2016) and, therefore, present a market segment that can help develop sustainable tourism (Šaparnienė et al., 2022). However, Vukic et al. (2015) show that youth preferences for traveling are not homogeneous. For example, Buffa (2015) notes that for young tourists, interest in suitability differs and affects their decision-making processes, motivations, and behaviors differently and should, therefore, be segmented.

The goal of market segmentation is to divide large, heterogeneous markets into smaller, more homogeneous units that are more accessible and whose needs can be better met. A market segment is, therefore, a group of people with similar needs, characteristics, and behavior, and which needs separate products or marketing strategies (Kotler et al., 1999). According to Tsiotsou and Vasioti (2006), segmentation can provide destinations with four benefits: It provides a base for target marketing, it helps in developing more effective marketing strategies, it facilitates destination differentiation, and it assists in identifying market opportunities and threats.

There are four major groups of characteristics that researchers can base their segmentation on, namely, demographic, geographic, behavioral, and psychographic. While demographic and geographic segmentation are still most commonly used in tourism, mainly because the information is easy to obtain, quite a few recent studies on winter tourism have focused on psychographic and behavioral segmentation (Dolnicar & Leisch, 2003; Füller & Matzler, 2008; Tsiotsou & Vasioti, 2006). Konu et al. (2011), for example, also used a behavioral approach by segmenting Finnish ski destination visitors based on attributes that affect ski destination choice. They conclude that ski destination choice attributes are an effective way to segment ski destination visitors and that further similar studies should be conducted to get a broader understanding of the topic.

The literature on segmentation of tourists with respect to their attitudes and behavior toward sustainability has also increased in the recent years. First of all, several authors note that there is still a gap in attitudes and behavior (see, e.g., Juvan & Dolnicar, 2014; Khanra et al., 2021), and the majority of the literature focus on the attitudes. Based on the extensive literature on environmentally friendly tourists, Dolnicar et al. (2008) suggest that people differ in their levels of sustainable tourist attitude. Based on tourists' attitude, Dolnicar (2004) segmented tourists visiting Austria into sustainable tourists and non-sustainable tourists, whereas Crouch et al. (2005) found that the two distinct groups of tourists – sustainable and non-sustainable tourists – differ significantly in sociodemographic and travel behavior and travel motivations.

For the young tourists, Buffa (2015) (following Weaver & Lawton, 2002) identified two groups of tourists: hard path young tourists and soft path young tourists, who differ with respect to the dimensions of sustainability and the influence this interest has on their decision-making processes, motivations, and behaviors. The group of hard path young tourists travels in small groups, enhances sustainability, are strongly environmentally commitment, and places emphasis on personal experiences. However, young tourists cannot be only distributed dichotomously into two groups. For example, McDonald et al. (2012) found there are so-called shades of green consumers who can to a greater or lesser extent commit to sustainable behavior. Kiatkawsin and Han (2017) applied the value–belief–norm theory and the expectancy theory to investigate the intention of young travelers to behave pro-environmentally and found that their pro-environmental behavior is influenced by values, beliefs, and norms to a different extent.

There are also two segmentation approaches: *a priori*, when the segmentation criterion is known in advance, and *a posteriori*, which is data-driven. It is not uncommon to combine these two approaches – to choose a particular segment of interest *a priori* and then use *a posteriori* segmentation to derive subgroups of that particular segment. A number of studies have already used this combined approach to further analyze young adults as an *a priori* segment (e.g., Pizam et al., 2004; Won et al., 2008; Won & Hwang, 2009), and while they deem the segmentation successful, they also welcome further investigation of this particular segment.

Research questions

On the basis of a thorough literature review, it has been established that the stake-holders at ski destinations could benefit from a deeper understanding of sustainable development of ski destinations from visitors' perspective. From the visitors' perspective, sustainable travel is essential for responsible consumption and production, and it is aligned with the SDGs (Seeler et al., 2021). While there is limited research on visitors' understanding of sustainable ski destinations, sustainability-aware tourists consider sustainability to be a crucial element in destination choice (Nowacki et al., 2021) and one that has been overlooked as well (Ashraf et al., 2020). As has been pointed out by Wehrli et al. (2012), the knowledge of different types of tourists regarding their attitude toward sustainability is important for destination managers since it helps understand how to approach potential customers of sustainable products. Especially for the young adults who are a significant demographic population in the travel industry and the market of the future, their attitudes toward sustainability are crucial. However, their preferences for sustainability differ, so it is necessary to segment them in order to understand their behavior better, the influence on decision-making, motivations, and behaviors (Buffa, 2015; McDonald et al., 2012; Šaparnienė et al., 2022). The goal of this study is to corroborate the findings of Konu et al. (2011) that segmentation based on ski destination attributes is effective and useful, and also to add new insights about the possible connection with tourists' attitudes toward sustainable development of ski destinations. To this end, the following research questions are formed:

RQ1: *Which segments of young adult ski destination visitors can be derived from elements that affect their decisions when choosing a ski destination?*

RQ2: *How do these segments differ in terms of the importance they give to different elements of sustainable development of ski destinations?*

Data and methodology

The items in the survey are based on an extensive literature review, which was focused on finding elements that affect visitors' decisions when choosing a ski destination and elements of sustainable development of ski destinations (Füller & Matzler, 2008; Konu et al., 2011; Tsiotsou & Vasioti, 2006; Unbehaun et al., 2008; Wehrli et al., 2012; Zehrer & Siller, 2007).

The questionnaire consisted of three parts. The first part related to the general characteristics of ski vacations (frequency of visiting ski resorts, company, type of accommodation, type of booking, and means of transport to ski resorts). The second part referred to the 16 elements of sustainable development that influence the decision to visit a specific ski destination (on a scale of 1–7) (e.g., good snow conditions, snow park, culinary offers, and additional services); 20 elements of sustainable development in a ski destination (on a scale of 1–7) (e.g., use of renewable energy in a destination, efficient waste management, and use of local products); and 10 elements of information about sustainable development of a ski destination (also on a scale of 1–7) (e.g., website of the destination and social media) as well as information about the willingness to pay a premium for a sustainable product or service and the willingness to travel longer for a sustainable ski destination. The third part contains standard demographic questions (age, gender, country of origin, marital status, annual income, etc.).

The research was conducted by using a web-based survey. The participants in this study were 267 young adults (aged 18–34 years), where 74% were aged 18–24 years and the rest 25–34 years (see Table 9.1). They came from EU member states, divided into four regions using the UNWTO classification: Central/Eastern (50), Northern (43), Western (113), and Southern/Mediterranean (61). The geographic composition of the sample was compared to that of the population, and despite the Western EU being slightly overrepresented, the chi-square test was non-significant, thus confirming sample representativeness in this regard. The male-to-female ratio also did not deviate from the population.

The data were analyzed in five stages. First, descriptive statistics were calculated in order to explore the sample characteristics that are presented in Table 9.1. The majority of respondents were single, and almost one-third lived with a partner. As for skiing, respondents spent an average of eight days per season skiing and were mostly with friends and/or family.

Next, a PCA with varimax rotation was performed in order to search for components of elements that affect the respondents' decision when choosing a ski

Table 9.1 Sample characteristics

Number of observations	267
Male (%)	52.81
Aged 18–24 years (%)	74.16
Region of residence (%)	
Central/Eastern	18.73
Northern	16.10
Western	42.32
Southern/Mediterranean	22.85
Family status (%)	
Living with partner	30.34
Living with partner and children	4.12
Single	64.79
Other	0.75
Ski holiday characteristics	
Average ski days per season	8.36
Company on ski vacations* (%)	
Family	59.55
Friends	82.77
Acquaintances	1.87
Coworkers	3.75
Alone	3.75
Other	1.87

* Multiple answers possible.

destination. PCA was chosen over factor analysis since the primary objective was not the search for underlying latent dimensions but rather to summarize the information in a fewer number of variables (Hair et al., 2010). Then, internal consistency of components was assessed using Cronbach's alpha, and summated scales for each component were calculated by averaging the corresponding variables.

Fourth, a K-means cluster analysis was performed on the components to search for homogeneous segments of respondents. This type of analysis combination (PCA and clustering) has been frequently used in studies of segmentation in tourism (e.g., Chang, 2006; Molera & Albaladejo, 2007; Park & Yoon, 2009; Wehrli et al., 2012). Furthermore, ANOVA was used to identify statistical differences between the clusters in terms of the components derived from the PCA. Finally, in the fifth stage, ANOVA was used to search for differences between the segments in terms of the importance they give to different elements of sustainable development of ski destinations. In addition, crosstabs with chi-square tests were used to search for possible differences regarding other (categorical) variables.

Results

The initial PCA of 16 items yielded a five-factor solution, but a number of variables had either loadings lower than 0.5 or cross-loadings higher than 0.4. In addition, the fourth and fifth components barely surpassed the eigenvalue of 1, thus implying a possible three-component solution. This was subsequently confirmed, and a final solution was obtained with 11 variables loading on three components: "Food & services," "Skiing & fun," and "Sustainability & nature" (Table 9.2). The loadings all exceeded 0.5, and there were no cross-loadings over 0.4.

Table 9.2 Components of elements that affect young adults' decision when choosing a ski destination

Variable	Food & services	Skiing & fun	Sustainability & nature
	Loadings		
Culinary at the hotel	0.885		
Culinary on the ski slope	0.876		
Culinary at the destination	0.848		
Additional services (wellness, spa, etc.)	0.699		
Good snow conditions		0.861	
Skiing and recreation		0.861	
Good ski slopes		0.786	
Fun		0.655	
Sustainable development of a resort			0.914
Sustainable development of a destination			0.907
Fresh air and nature			0.613
Variance explained (%)	25.8	23.6	19.7
Cronbach's alpha	0.859	0.806	0.776

The eigenvalues surpassed 1, and the total explained variance amounted to 69%. The Keiser–Meyer–Olkin measure of sampling adequacy (KMO = 0.743) and Bartlett's test of sphericity ($p < 0.001$) both confirmed a satisfactory solution. Cronbach's alpha values were 0.859, 0.806, and 0.776, which indicate acceptable levels of reliability (Hair et al., 2010).

The components were then used as summated scales in a K-means cluster analysis. Since the K-means is a nonhierarchical clustering algorithm, the number of clusters has to be specified in advance. Therefore, several different cluster solutions (ranging from 3 to 7 clusters) were obtained, and after examination and comparison, the six-cluster solution was chosen to be the most meaningful and interpretable (Table 9.3.). Results show that clusters indeed possess different characteristics, which can be seen by comparing the cluster component means to those of the whole sample. The first cluster sees skiing as important but feels the opposite about sustainability; therefore, we call this cluster *anti-sustainability skiers*. The second cluster is diametrically opposite to the first, so it is appropriately named *anti-skiing sustainabilitists*. The third cluster finds both skiing and sustainability important; therefore, they are called *sustainable skiers*. In comparison, the fifth cluster finds food more important and skiing not as much; thus, we named them *sustainable hedonists*. The fourth cluster sees skiing as the only important aspect while completely ignoring food and services – the appropriate name for them would be *solely skiers*. Lastly, the sixth cluster does not find any of the elements important, and what is most interesting, they care about skiing the least. These are deemed *indifferent anti-skiers*.

After naming the clusters, chi-square tests and ANOVA were used to further examine the potential differences between the clusters regarding other variables. In terms of gender, region, and family status, the chi-square test was not significant. However, some interesting patterns are still visible. Cluster 6 (*indifferent anti-skiers*) are predominantly male, single, and almost exclusively Western and

Table 9.3 Final cluster centers

Component	Whole sample	Cluster number						ANOVA	
		1	*2*	*3*	*4*	*5*	*6*	*F*	*p*
		Anti-sustain-ability skiers	Anti-skiing sustain-abilitists	Sust-ainable skiers	Solely skiers	Sustain-able hedonists	Indiffe-rent anti-skiers		
Food & services	3.93	3.88	3.77	4.15	1.62	5.81	3.75	179.06	<0.001
Skiing & fun	5.66	6.05	4.47	6.24	6.03	5.84	1.85	125.84	<0.001
Sustainability & nature	4.50	3.33	5.08	5.51	3.90	5.16	3.10	62.24	<0.001
Cluster size (%)		22.8	12.0	23.6	16.9	19.9	4.9		

Southern European, while Cluster 3 (*sustainable skiers*) has a higher proportion of women, people with partners, and more Northern and Eastern Europeans (see Table 9.4.). In terms of the importance of different elements of sustainable development of ski destinations, ANOVA shows significant differences between the clusters regarding all 20 elements. Although not all pairwise comparisons are statistically significant, upon closer examination an obvious pattern emerges. Compared to other clusters, *indifferent anti-skiers* give the least importance to nearly all elements, with *anti-sustainability skiers* and *solely skiers* being generally not much better. On the other hand, *sustainable skiers* are almost exclusively the ones that assign the most importance, with *sustainable hedonists* in close second place and *anti-skiing sustainabilitists*, therefore, left in the middle.

Table 9.4 Cluster characteristics

	Anti-sustainability skiers	*Anti-skiing sustainabilitists*	*Sustainable skiers*	*Solely skiers*	*Sustainable hedonists*	*Indifferent anti-skiers*
Male (%)	52.46	56.25	46.03	57.78	52.83	61.54
Aged 18–24 years (%)	83.61	65.63	76.19	64.44	75.47	69.23
Region of residence (%)						
Central/Eastern	18.03	21.88	23.81	13.33	18.87	7.69
Northern	18.03	15.63	20.63	13.33	15.09	0.00
Western	40.98	34.38	41.27	55.56	35.85	53.85
Southern/Mediterranean	22.95	28.13	14.29	17.78	30.19	38.46
Family status (%)						
Living with partner	31.15	28.13	38.10	33.33	22.64	15.38
Living with partner and children	1.64	9.38	4.76	2.22	5.66	0.00
Single	67.21	62.50	55.56	64.44	69.81	84.62
Other	0.00	0.00	1.59	0.00	1.89	0.00
Ski holiday characteristics						
Average ski days per season	10.26	4.84	9.06	5.49	7.49	18.23
Company on ski vacations* (%)						
Family	70.49	43.75	52.38	57.78	66.04	61.54
Friends	85.25	84.38	87.30	80.00	77.36	76.92
Acquaintances	1.64	0.00	3.17	0.00	3.77	0.00
Coworkers	6.56	0.00	3.17	0.00	7.55	0.00
Alone	3.28	3.13	3.17	11.11	0.00	0.00
Other	1.64	0.00	3.17	0.00	3.77	0.00
Preferred way of booking * (%)						
Accommodation provider	75.41	75.00	77.78	66.67	69.81	53.85
Local tourist information office	27.87	34.38	25.40	24.44	43.40	61.54
Tourist agency	29.51	18.75	22.22	15.56	28.30	23.08

* Multiple answers possible.

Discussion, conclusions, and implications for practice

Based on a literature review, a set of ski destination attributes was developed for the purpose of examining the importance of these attributes to young adults when choosing a ski destination. Also, a set of elements of sustainable development of ski destinations was developed in order to investigate their relative importance to different young adult ski destination visitors. First, a PCA was performed on the ski destination attributes, which produced three components: "Food & services," "Skiing & fun," and "Sustainability & nature." Summated scales were computed for each component and used in a K-means cluster analysis, which subsequently produced a satisfactory six-cluster solution. The resulting clusters (segments) were then named – *anti-sustainability skiers, anti-skiing sustainabilitists, sustainable skiers, sustainable hedonists, solely skiers,* and *indifferent anti-skiers* – and compared regarding other characteristics, namely, demographics and the elements of sustainable development of ski destinations, by using chi-square tests and ANOVA.

The results show that the differences between the segments regarding the importance of the elements of sustainable development of ski destinations were all statistically significant – the least sustainable segment was *indifferent anti-skiers*, whereas *sustainable skiers* were the most sustainable segment. The chi-square tests on demographics were not significant; however, the data seem to imply that *indifferent anti-skiers* consist of more males, singles, and Western and Southern Europeans, while *sustainable skiers* are composed of more women, more people with partners, and more Northern and Eastern Europeans. These differences should be investigated further, and these results could provide a good starting point for future studies. Although aimed at having a representable geographic composition of the sample, the sample data and the analysis have some limitations common to the chosen sampling method (e.g., sampling bias). In addition, we also suggest a more longitudinal approach in the future to investigate possible changes in the population's attitudes toward sustainability elements of ski destinations and also investigate if the time and effort devoted to educating the young adults have changed the size of the clusters.

Tourism organizations and destination management organizations should understand that young adults have different attitudes toward sustainability. And this study shows that elements that influence young adults' decisions to choose a ski destination are an effective way to segment young adult ski destination visitors. Furthermore, the results show that segments of young adult ski destination visitors also differ in terms of the importance they place on different elements of sustainable development of ski destinations. This knowledge can be used by destination managers in developing new sustainable products and services in order to better meet the needs and expectations of visitors. In addition, segmentation of young adult ski visitors can be used to target specific segments when different sustainable products and services are offered. For example, *sustainability skiers* are mostly from Western Europe, younger, single or living with a partner, and spend on average about nine ski days per ski season. They usually spend their ski days with friends and book their accommodation directly with the accommodation provider. The latter is important for the question of where information on sustainable products and services should be provided. In contrast, *indifferent anti-skiers* book their ski trips through local tourist information centers, and *sustainable hedonists* do so to a large extent. This knowledge can also be used to target specific segments of young visitors to ski destinations.

The results also have theoretical implications, since this study contributes to a relatively unexplored topic of ski destination visitors' attitudes toward different elements of sustainable development. It also provides additional insight into young adults' preferences when choosing a ski destination and can also provide a basis for future research of sustainability at ski destinations from the visitors' perspective.

References

Ashraf, M. S., Hou, F., Kim, W. G., Ahmad, W., & Ashraf, R. U. (2020). Modeling tourists' visiting intentions toward ecofriendly destinations: Implications for sustainable tourism operators. *Business Strategy and the Environment*, 29(1), 54–71. https://doi.org/10.1002/bse.2350

Bank, M., & Wiesner, R. (2011). Determinants of weather derivatives usage in the Austrian winter tourism industry. *Tourism Management*, 32(1), 62–68.

Bausch, T., & Gartner, W. C. (2020). Winter tourism in the European Alps: Is a new paradigm needed? *Journal of Outdoor Recreation and Tourism*, 31, 100297. https://doi.org/10.1016/j.jort.2020.100297

Buckley, R. (2012). Sustainable tourism: Research and reality. *Annals of Tourism Research*, 39(2), 528–546. https://doi.org/10.1016/j.annals.2012.02.003

Buffa, F. (2015). Young tourists and sustainability. Profiles, attitudes, and implications for destination strategies. *Sustainability*, 7(10), 14042–14062. https://doi.org/10.3390/su71014042

Chang, J. (2006). Segmenting tourists to aboriginal cultural festivals: An example in the Rukai tribal area, Taiwan. *Tourism Management*, 27(6), 1224–1234.

Crouch, G., Devinney, T., Dolnicar, S., Huybers, T., Louviere, J., & Oppewal, H. (2005). New horses for old courses: Questioning the limitations of sustainable tourism to supply-driven measures and the nature-based context. *Faculty of Commerce – Papers (Archive)*. Retrieved from https://ro.uow.edu.au/commpapers/70

Dolnicar, S. (2004). Insights into sustainable tourists in Austria: A data-based a priori segmentation approach. *Journal of Sustainable Tourism*, 12(3), 209–218. https://doi.org/10.1080/09669580408667234

Dolnicar, S. (2006). Nature-conserving tourists: The need for a broader perspective. *Anatolia*, 17(2), 235–255. https://doi.org/10.1080/13032917.2006.9687188

Dolnicar, S., & Leisch, F. (2003). Winter tourist segments in Austria: Identifying stable vacation styles using bagged clustering techniques. *Journal of Travel Research*, 41(3), 281–292.

Dolnicar, S., & Leisch, F. (2008). An investigation of tourists' patterns of obligation to protect the environment. *Journal of Travel Research*, 46(4), 381–391. https://doi.org/10.1177/0047287507308330

Dolnicar, S., Crouch, G. I., & Long, P. (2008). Environment-friendly tourists: What do we really know about them? *Journal of Sustainable Tourism*, 16(2), 197–210. https://doi.org/10.2167/jost738.0

Dragović, N., & Pašić, M. (2020). Sustainable tourism in ski resorts of Europe and the world. *Tourism and Sustainable Development Challenges, Opportunities, and Contradictions*. https://doi.org/10.15308/Sitcon-2020–108–116

Flagestad, A., & Hope, C. A. (2001). Strategic success in winter sports destinations: A sustainable value creation perspective. *Tourism Management*, 22(5), 445–461.

Fredman, P., & Margaryan, L. (2021). 20 years of Nordic nature-based tourism research: A review and future research agenda. *Scandinavian Journal of Hospitality and Tourism*, 21(1), 14–25. https://doi.org/10.1080/15022250.2020.1823247

Füller, J., & Matzler, K. (2008). Customer delight and market segmentation: An application of the three-factor theory of customer satisfaction on life style groups. *Tourism Management*, 29(1), 116–126.

Gilg, A., Barr, S., & Ford, N. (2005). Green consumption or sustainable lifestyles? Identifying the sustainable consumer. *Futures*, 37(6), 481–504.

Hair, J. F., Black, W. C., Babin, B. J., & Anderson, R. E. (2010). *Multivariate data analysis* (7th ed.). Upper Saddle River, New Jersey: Prentice Hall.

Hill, J., & Lee, H.-H. (2012). Young generation Y consumers' perceptions of sustainability in the apparel industry. *Journal of Fashion Marketing and Management*, 16(4), 477–491.

Juvan, E., & Dolnicar, S. (2014). The attitude–behaviour gap in sustainable tourism. *Annals of Tourism Research*, 48, 76–95. https://doi.org/10.1016/j.annals.2014.05.012

Kastenholz, E., Eusébio, C., & Carneiro, M. J. (2018). Segmenting the rural tourist market by sustainable travel behaviour: Insights from village visitors in Portugal. *Journal of Destination Marketing & Management*, 10, 132–142. https://doi.org/10.1016/j.jdmm.2018.09.001

Khanra, S., Dhir, A., Kaur, P., & Mäntymäki, M. (2021). Bibliometric analysis and literature review of ecotourism: Toward sustainable development. *Tourism Management Perspectives*, 37, 100777. https://doi.org/10.1016/j.tmp.2020.100777

Kiatkawsin, K., & Han, H. (2017). Young travelers' intention to behave pro-environmentally: Merging the value-belief-norm theory and the expectancy theory. *Tourism Management*, 59, 76–88. https://doi.org/10.1016/j.tourman.2016.06.018

Knowles, N. (2019). Can the North American ski industry attain climate resiliency? A modified Delphi survey on transformations towards sustainable tourism. *Journal of Sustainable Tourism*, 27(3), 380–397. https://doi.org/10.1080/09669582.2019.1585440

Konu, H., Laukkanen, T., & Komppula, R. (2011). Using ski destination choice criteria to segment Finnish ski resort customers. *Tourism Management*, 32(5), 1096–1105.

Kotler, P., Armstrong, G., Saunders, J., & Wong, V. (1999). *Principles of marketing* (2nd ed.). London: Prentice Hall Europe.

Luthe, T., & Schläpfer, F. (2011). Effects of third-party information on the demand for more sustainable consumption: A choice experiment on the transition of winter tourism. *Environmental Innovation and Societal Transitions*, 1(2), 234–254.

Mcdonald, S., Oates, C. J., Alevizou, P. J., Young, C. W., & Hwang, K. (2012). Individual strategies for sustainable consumption. *Journal of Marketing Management*, 28, 445–68.

Miller, G., Rathouse, K., Scarles, C., Holmes, K., & Tribe, J. (2010). Public understanding of sustainable tourism. *Annals of Tourism Research*, 37(3), 627–645.

Molera, L., & Albaladejo, I. P. (2007). Profiling segments of tourists in rural areas of South-Eastern Spain. *Tourism Management*, 28(3), 757–767.

Nielsen. (2017). *Young and ready to travel: A look at millennial travelers.* Retrieved from www.nielsen.com/wp-content/uploads/sites/3/2019/04/nielsen-millennial-traveler-study-jan-2017.pdf

Nordin, S., Volgger, M., Gill, A., & Pechlaner, H. (2019). Destination governance transitions in skiing destinations: A perspective on resortisation. *Tourism Management Perspectives*, 31, 24–37. https://doi.org/10.1016/j.tmp.2019.03.003

Nowacki, M., Chawla, Y., & Kowalczyk-Anioł, J. (2021). What drives the eco-friendly tourist destination choice? The Indian perspective. *Energies*, 14(19), 6237. https://doi.org/10.3390/en14196237

OECD. (2018). *OECD tourism trends and policies 2018.* OECD Publishing, Paris. https://doi.org/10.1787/tour-2018-en

Park, D.-B., & Yoon, Y.-S. (2009). Segmentation by motivation in rural tourism: A Korean case study. *Tourism Management*, 30(1), 99–108.

Pizam, A., Jeong, G., Reichel, A., Van Boemmel, H., Lusson, J. M., Steynberg, L., . . . & Montmany, N. (2004). The relationship between risk-taking, sensation-seeking, and the tourist behavior of young adults: A cross-cultural study. *Journal of Travel Research*, 42(3), 251–260.

Šaparnienė, D., Mejerė, O., Raišutienė, J., Juknevičienė, V., & Rupulevičienė, R. (2022). Expression of behavior and attitudes toward sustainable tourism in the youth population: A search for statistical types. *Sustainability*, 14(1), 473. https://doi.org/10.3390/su14010473

Seeler, S., Zacher, D., Pechlaner, H., & Thees, H. (2021). Tourists as reflexive agents of change: Proposing a conceptual framework towards sustainable consumption. *Scandinavian Journal of Hospitality and Tourism*, 21(5), 567–585. https://doi.org/10.1080/15022250.2021.1974543

Škare, M., Soriano, D. R., & Porada-Rochoń, M. (2021). Impact of COVID-19 on the travel and tourism industry. *Technological Forecasting and Social Change*, 163, 120469. https://doi.org/10.1016/j.techfore.2020.120469

Škorić, S. (2008). The implementation of the sustainable development principles in winter sports tourism. In *4th International Conference: An enterprise Odyssey: Tourism – governance and entrepreneurship proceedings*. Zagreb: University of Zagreb.

Steiger, R. (2012). Scenarios for skiing tourism in Austria: Integrating demographics with an analysis of climate change. *Journal of Sustainable Tourism*, 20(6), 867–882. https://doi.org/10.1080/09669582.2012.680464

Steiger, R., Knowles, N., Pöll, K., & Rutty, M. (2022). Impacts of climate change on mountain tourism: A review. *Journal of Sustainable Tourism*, 0(0), 1–34. https://doi.org/10.1080/09669582.2022.2112204

Tsiotsou, R., & Vasioti, E. (2006). Satisfaction: A segmentation criterion for "Short Term" visitors of mountainous destinations. *Journal of Travel & Tourism Marketing*, 20(1), 61–73.

UN. (2019). Sustainable development goals. *United Nations*. Retrieved from www.un.org/sustainabledevelopment/sustainable-development-goals/

Unbehaun, W., Pröbstl, U., & Haider, W. (2008). Trends in winter sport tourism: Challenges for the future. *Tourism Review*, 63(1), 36–47.

Verbeek, D. H. P., Bargeman, A., & Mommaas, J. T. (2011). A sustainable tourism mobility passage. *Tourism Review*, 66(4), 45–53.

Vukic, M., Kuzmanovic, M., & Kostic Stankovic, M. (2015). Understanding the heterogeneity of generation Y's preferences for travelling: A conjoint analysis approach. *International Journal of Tourism Research*, 17(5), 482–491. https://doi.org/10.1002/jtr.2015

Weaver, D. B., & Lawton, L. J. (2002). Overnight ecotourist market segmentation in the Gold Coast hinterland of Australia. *Journal of Travel Research*, 40(3), 270–280. https://doi.org/10.1177/004728750204000305

Wehrli, R., Egli, H., Lutzenberger, M., Pfister, D., & Stettler, J. (2012). Tourists' understanding of sustainable tourism: An analysis in eight countries. *GSTF Business Review (GBR)*, 2(2), 219–224.

Won, D., & Hwang, S. (2009). Factors influencing the college skiers and snowboarders' choice of a ski destination in Korea: A conjoint study. *Managing Leisure*, 14(1), 17–27.

Won, D., Bang, H., & Shonk, D. J. (2008). Relative importance of factors involved in choosing a regional ski destination: Influence of consumption situation and recreation specialization. *Journal of Sport & Tourism*, 13(4), 249–271.

Wondirad, A. (2019). Does ecotourism contribute to sustainable destination development, or is it just a marketing hoax? Analyzing twenty-five years contested journey of ecotourism through a meta-analysis of tourism journal publications. *Asia Pacific Journal of Tourism Research*, 24(11), 1047–1065. https://doi.org/10.1080/10941665.2019.1665557

World Tourism Organization. (2011). *Affiliate members global report, volume 2 – The power of youth travel*, UNWTO, Madrid. https://doi.org/10.18111/9789284414574

World Tourism Organization. (2016). *Affiliate members global report, volume 2 – The power of youth travel*, UNWTO, Madrid. https://doi.org/10.18111/9789284417162

Zehrer, A., & Siller, H. (2007). Destination goods as travel motives – the case of the Tirol. *Tourism Review*, 62(3–4), 39–46.

10 Stakeholder sentiment of SDGs of Beijing Olympics 2020

Lori Pennington-Gray, Seonjin Lee,
and Khalid Ballouli

Introduction

The Olympic Games, as a global mega event, have the potential to influence public sentiment and shape perceptions on various issues, including sustainable development. The Sustainable Development Goals (SDGs) provide a comprehensive framework for addressing global challenges related to poverty, inequality, climate change, and more. The achievement of the Millennium Development Goals (MDGs) can have an influence on public sentiment regarding the Olympics. The MDGs, established by the United Nations in 2000, were a set of eight development goals aimed at addressing global challenges such as poverty, education, health, and environmental sustainability. While the MDGs have since expired and been replaced by the SDGs, their impact on public sentiment during the period of their implementation can be considered.

Public sentiment at the Olympics is influenced by the perception of whether the games contribute to the achievement of the MDGs. If the Olympics are seen as promoting social progress, such as poverty reduction, education accessibility, gender equality, and improved health outcomes, it is likely to generate positive sentiment. The adoption of sustainable practices and the incorporation of social and environmental considerations throughout the games can be perceived as a reflection of the event's commitment to global development. For instance, the Olympic Agenda 2020 strategy emphasizes sustainability principles, inspiring hope and confidence that the games are contributing to a better world (IOC, n.d.). The alignment of the games with the MDGs can foster a sense of purpose and global cooperation, leading to a favorable public perception.

Recognizing the significant influence and reach of the Olympic Games, the IOC has developed the IOC Sustainability Strategy to ensure that the Olympic Movement contributes to sustainable development on a global scale. Under the IOC Sustainability Strategy, 11 SDGs have been identified as pertinent to the Olympic Movement. These SDGs include SDG 3 (Good health and well-being), SDG 4 (Quality education), SDG 5 (Gender equality), SDG 6 (Clean water and sanitation), SDG 7 (Affordable and clean energy), SDG 8 (Decent work and economic growth), SDG 9 (Industry, innovation, and infrastructure), SDG 11 (Sustainable cities and communities), SDG 12 (Responsible consumption and production),

DOI: 10.4324/9781003384786-10

SDG 13 (Climate action), and SDG 17 (Partnerships for the goals) (Schulenkorf, 2012).While these are some of the key SDGs often associated with the Olympics, it's important to note that the connection between the SDGs and the Olympics may vary depending on the specific context, host city, or organizing committee's priorities. The alignment with the SDGs can evolve over time as sustainability efforts and social priorities advance. The perception that the Olympics are actively contributing to addressing climate change and environmental challenges can boost public sentiment.

Several contributions have been made by the Olympic Games toward achieving SDGs. First, the Olympic Games have actively contributed to addressing climate change through various initiatives. Efforts such as carbon offsetting, renewable energy integration, sustainable infrastructure, and awareness campaigns have been implemented to reduce the ecological footprint of the games. Second, an additional approach is carbon offsetting, where the Olympics invest in projects that compensate for the emissions generated during the event. For example, the London 2012 Olympics aimed to be the first carbon-neutral games by offsetting residual emissions through renewable energy projects (International Olympic Committee, 2012).

Third, the integration of renewable energy sources has become a key focus for the Olympics. The Tokyo 2020 Olympics, for instance, aimed to power the games with 100% renewable energy. This involved the installation of solar panels in venues and the use of hydrogen as an alternative fuel source (Tokyo, 2020, 2021).

Fourth, sustainable infrastructure development has also been emphasized. Host cities are encouraged to design energy-efficient venues, utilize sustainable materials, and implement effective water and waste management strategies aligned with sustainability principles.

In addition to these operational measures, the Olympic Games serve as a platform for raising awareness about climate change and promoting sustainable practices. Educational campaigns, sustainable legacy initiatives, and efforts to encourage behavioral changes among athletes, spectators, and local communities contribute to this aspect.

Moreover, the social legacy left by the Olympics in relation to the MDGs can shape public sentiment. If the games result in lasting improvements in infrastructure, education, healthcare, and social inclusion, they can be seen as positive contributions to the pursuit of the SDGs. Conversely, if the Olympics are perceived as failing to deliver substantial social benefits or exacerbating inequalities, public sentiment may be more critical or negative.

Literature review

The role of sports in achieving SDGs

Sports have emerged as a powerful tool in promoting sustainable development and peace. Since the inception of the MDGs in 2000, sports have played a significant role in advancing each of the eight goals outlined in the Millennium Declaration

(Manzenreiter, 2005). The United Nations, in its 2015 resolution 70/1 "Transforming our world: A 2030 Agenda for Sustainable Development," reaffirmed the importance of sports in driving social progress. Sports contribute to sustainable development by fostering tolerance, respect, and empowerment, particularly among women, youth, individuals, and communities. It also helps achieve health, education, and social integration objectives.

The United Nations Office on Sport for Development and Peace (UNOSDP) has long recognized the potential of sports to promote development and has actively supported its utilization in various initiatives, from large-scale sports projects to community-based competitions (Manzenreiter, 2005). These initiatives harness the power of sports to its fullest extent in advancing development goals. Sports not only have a direct impact on physical health, but they also instill the importance of leading a healthy lifestyle among children and adolescents, promoting physical activity and combating noncommunicable diseases. Furthermore, research conducted by the World Health Organization suggests that exercise contributes to mental well-being, boosting self-esteem and positively impacting individuals with depression and anxiety symptoms.

Sport for Sustainable Development programs integrate sports with non-sport components to enhance their impact on local, regional, national, and transnational sustainability initiatives. These programs aim to create mutually reinforcing efforts by collaborating, forming partnerships, and taking coordinated action. Such programs not only benefit the target groups but also empower participants at every stage of delivering sustainable activities (Gasser & Levinsen, 2004). By leveraging the power of sports and incorporating sustainability principles, these initiatives address social, economic, and environmental challenges. They promote health and well-being, provide educational opportunities, foster gender equality and inclusivity, reduce inequalities, enhance urban sustainability, and promote responsible consumption.

Overall, sports serve as a catalyst for achieving the SDGs, extending its impact beyond the boundaries of sports itself. Through collaboration, innovation, and holistic approaches, sports can contribute significantly to sustainable development efforts worldwide. In conclusion, sports has a multifaceted role in achieving the SDGs. It promotes physical and mental well-being, empowers women, breaks down stereotypes, and fosters social cohesion and peace. Harnessing the power of sports can unlock its full potential in advancing sustainable development objectives and creating a more inclusive and peaceful world.

The conceptual framework

The article titled "A systematic review of the role of sport in achieving the Sustainable Development Goals" published in the journal *Nature Sustainability* by Muller et al. (2021) explored the relationship between sports and the SDGs. The study provided a comprehensive review of existing literature and research on how sports can contribute to achieving the SDGs. The authors highlighted that sports have the potential to positively impact various SDGs, including health and well-being, education, gender equality, climate action, sustainable cities and communities, and more. They emphasized that sports can be a powerful tool for promoting social change, fostering inclusivity, and addressing global challenges.

The article identified key mechanisms through which sports can contribute to the SDGs. These include promoting physical activity and health, providing educational opportunities, supporting social cohesion, fostering gender equality and empowerment, promoting environmental sustainability, and stimulating economic development.

Also, the authors acknowledged the need for further research and evaluation to better understand the effectiveness and long-term impacts of sport-based interventions in achieving the SDGs. Their study emphasized the importance of collaboration between sports organizations, policymakers, and researchers to maximize the potential of sports in contributing to sustainable development.

The conceptual model presented in Figure 10.1 provides a framework for understanding the sustainability of the Olympic Games. It takes into account the ongoing

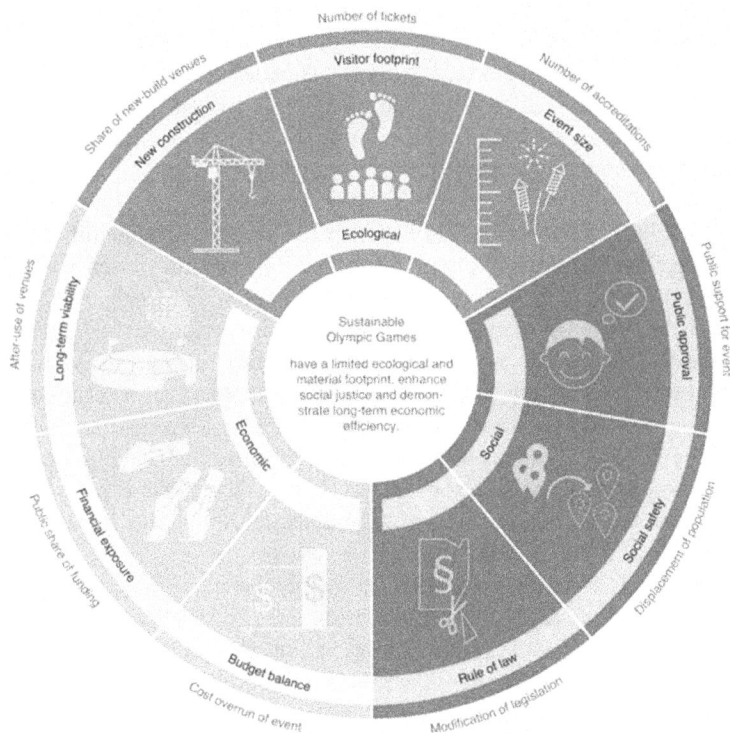

Figure 10.1 Definition and conceptual model of sustainability in the Olympic Games.

Source: Müller, Martin & Wolfe, Sven & Gaffney, Christopher & Gogishvili, David & Leick, Annick & Hug, Miriam (2021)

discourse and affect on sustainability, which emphasizes minimizing resource use while ensuring social and economic well-being. By incorporating these dimensions, the model aligns with policy discussions on sustainability, such as the United Nations' 17 SDGs that aim for equitable human development while reducing resource consumption, as well as the Paris Agreement.

Affect refers to the broad spectrum of emotional and evaluative experiences, including positive and negative feelings. In recent years, there has been a global shift toward greater environmental consciousness. As the world's premier sporting event, the Olympics have been a unique platform for promoting, critiquing, and judging the games' commitment to sustainability. This expression of sentiment reflects various publics affect toward sustainability.

Theory of affect

The theory of affect proposes that sentiments are a manifestation of affect, which encompasses emotions, moods, and attitudes. According to this theory, sentiments are subjective experiences and evaluations that individuals express based on their affective states. Affect refers to the broad spectrum of emotional and evaluative experiences, including positive and negative feelings.

The theory suggests that individuals' affective states influence their expressions of sentiment. For example, if a person is experiencing positive affect, they are more likely to express positive sentiments, while negative affect may lead to the expression of negative sentiments. Affect can be influenced by various factors, such as personal experiences, cognitive appraisals, social interactions, and cultural norms.

Understanding the theory of affect helps in interpreting sentiment analysis results by recognizing the underlying emotional states that drive sentiment expressions. By considering the affective component, sentiment analysis can provide insights into individuals' attitudes, preferences, and reactions.

The theory of affect is highly relevant to sentiment analysis as it provides insights into the underlying emotional states that influence sentiment expressions. Sentiment analysis aims to analyze and understand the sentiment or opinion expressed in text data, such as social media posts, customer reviews, or news articles. By incorporating the theory of affect, sentiment analysis can go beyond the surface-level polarity (positive, negative, and neutral) and delve into the emotions and attitudes that drive sentiment.

The theory of affect suggests that sentiments are closely tied to individuals' affective states, which encompass emotions, moods, and attitudes. Sentiment analysis algorithms can leverage this understanding by analyzing linguistic cues, such as specific words, phrases, or sentiment indicators, to infer the emotional content and intensity of sentiment expressions.

For example, sentiment analysis can distinguish between different emotional states like joy, anger, sadness, or surprise. By identifying and categorizing the affective dimensions of sentiment, sentiment analysis can provide more nuanced insights into the attitudes, preferences, and reactions of individuals or groups.

Furthermore, incorporating the theory of affect in sentiment analysis can enhance the accuracy and reliability of sentiment classification models. By considering the emotional context, sentiment analysis algorithms can better understand the underlying sentiment expressed in the text, even in cases where the sentiment might be more subtle or ambiguous.

Sentiment analysis as a method to measure public approval

Social sentiment analysis plays a pivotal role in measuring public support by providing valuable insights into the opinions, emotions, and attitudes expressed by individuals on social media platforms. It leverages natural language processing and machine learning techniques to analyze and classify social media posts, comments, and mentions, enabling organizations to understand public sentiment toward specific topics or events.

Measuring public support is crucial for organizations, brands, and governments to gauge public perception and make informed decisions. Social sentiment analysis offers real-time insights into public sentiment, allowing for timely response and adaptation to changing opinions (Agarwal et al., 2011). It helps identify key issues and concerns raised by the public, enabling effective problem-solving and issue resolution (Thelwall et al., 2012). By evaluating the impact of campaigns or announcements on public sentiment, organizations can assess the effectiveness of their initiatives (Reyes-Menendez et. al, 2018).

Negative sentiment waning over time is a phenomenon that has been observed in various contexts and is supported by research studies. It suggests that negative sentiment expressed by individuals tends to diminish or decrease as time progresses.

One possible explanation for this phenomenon is the natural process of emotional recovery or adaptation. Over time, individuals may become less affected or emotionally charged by certain events or issues, leading to a decrease in negative sentiment. This can be attributed to psychological factors such as habituation, resilience, or cognitive coping mechanisms (Kahneman et al., 1993).

Furthermore, negative sentiment may also decline due to the influence of social dynamics and collective behavior. As people interact and engage in discussions or discourse about a particular topic, they may reach a point of saturation or exhaustion, leading to a decrease in the expression of negative sentiment. This can be observed in online communities or social media platforms, where initial intense reactions tend to subside as discussions progress (Tumasjan et al., 2010). It is important to note that the decline of negative sentiment over time does not necessarily indicate a resolution of the underlying issues or a change in opinions. It simply reflects a shift in the intensity of emotional expression related to the topic or event.

Positive sentiment over time refers to the phenomenon where individuals' positive attitudes, emotions, or opinions regarding a particular topic or event remain consistent or even increase as time progresses. While it is commonly observed that negative sentiment may wane over time, positive sentiment can exhibit different patterns.

One possible explanation for positive sentiment remaining stable or increasing over time is the impact of positive experiences and continued positive reinforcement. When individuals continually encounter positive outcomes, achievements, or improvements related to a specific topic, their positive sentiment may persist or even strengthen. This can be seen in situations where ongoing progress, advancements, or success stories contribute to a sustained positive outlook.

Additionally, the influence of social factors and social dynamics can contribute to the maintenance or growth of positive sentiment over time. When individuals observe and engage with others who express and reinforce positive views, attitudes, or experiences related to a particular topic, it can create a ripple effect, leading to the sustained or amplified expression of positive sentiment within a community or society.

Furthermore, the presence of ongoing positive communication, campaigns, or initiatives focused on a specific topic can contribute to the long-term maintenance or increase of positive sentiment. Consistent exposure to messages, stories, or events that highlight the benefits, achievements, or positive impacts associated with the topic can reinforce positive sentiment among individuals and the wider public.

It is important to note that while positive sentiment can endure or grow over time, it may also be influenced by contextual factors, individual experiences, and external events. The complexity of human emotions and the interplay of various factors make it crucial to consider a range of elements when studying and understanding the trajectory of positive sentiment over time.

Social sentiment analysis also enables organizations to gain a broader understanding of public sentiment by segmenting the data based on demographics, locations, or other relevant factors (Agarwal et al., 2011). This allows for targeted analysis and customized strategies tailored to specific audience segments. Moreover, integrating sentiment analysis with other data sources, such as surveys or focus groups, provides a more comprehensive view of public support.

While social sentiment analysis provides valuable insights, it does have limitations. It may not capture the sentiment of the entire population, as not everyone expresses their opinions publicly or participates in social media discussions. This can lead to a potential sampling bias and limit the generalizability of the findings to the broader population. Additionally, algorithms used in sentiment analysis may have biases and may struggle with detecting sarcasm, irony, or nuanced expressions. Additionally, language variations, slang, and abbreviations commonly used on social media platforms can pose difficulties in correctly interpreting the sentiment of the tweets.

Nevertheless, social sentiment analysis remains a powerful tool for measuring public support, aiding organizations in understanding public sentiment, making data-driven decisions, and adapting their strategies accordingly.

Stakeholder demand for SDGs in the Olympics

Sporting events have emerged as influential platforms for promoting sustainability and addressing global challenges. The demand for integrating the SDGs into sports has been on the rise, driven by various stakeholders and factors. Athlete activism, fan consciousness, sponsorship interests, and the recognition of sports organizations' role in sustainable development have all contributed to this growing demand. As the world becomes more attuned to environmental and social issues, sporting events are increasingly expected to align with the SDGs and demonstrate a commitment to sustainable practices. This trend reflects a broader shift in the sports industry toward leveraging its immense popularity and reach to drive positive change.

The demand for integrating SDGs into the Olympics has been steadily increasing. Various stakeholders, including athletes, fans, sponsors, and sports organizations, have recognized the importance of aligning the Olympics with sustainable development principles. Athletes have emerged as influential advocates for sustainability and social issues within the sports community. They use their platforms to raise awareness about environmental conservation, social justice, and human rights. Athlete activism has sparked a demand for the Olympics to prioritize sustainability and actively contribute to addressing global challenges (UNEP, 2020).

Fans, as a crucial audience, have also become more conscious of sustainability and social responsibility. They expect sporting events, including the Olympics, to align with the SDGs and demonstrate a commitment to environmental stewardship, inclusivity, and ethical practices. Fan engagement and support are vital for the success and reputation of the Olympics, driving the demand for sustainable development integration (Wylleman et al., 2019).

Sponsors and brands recognize the value of associating their names with events that prioritize sustainability. They seek partnerships with the Olympics to enhance their corporate social responsibility image and promote sustainable practices. Sponsors often require the Olympics to align with the SDGs as part of their commitment to sustainability (Beek et. al., 2023).

Sports organizations and governing bodies have acknowledged their responsibility to contribute to sustainable development. They understand the significance of integrating the SDGs into the Olympics to fulfill their corporate social responsibility goals. By incorporating sustainable practices, promoting inclusivity, and supporting community development, sports organizations enhance their reputation and demonstrate their commitment to social and environmental issues (UNEP, 2020).

In conclusion, the demand for integrating the SDGs into the Olympics is driven by the collective efforts of athletes, fans, sponsors, and sports organizations. Athlete activism has brought issues of sustainability and social justice to the forefront, inspiring the call for the Olympics to prioritize these principles. Fans, as a crucial audience, expect the Olympics to align with the SDGs, reflecting their growing consciousness of sustainability and social responsibility. Sponsors and brands recognize the value of associating their names with sustainable events and require the Olympics to demonstrate a commitment to the SDGs. Sports organizations and governing bodies understand their responsibility to contribute to sustainable development and actively incorporate the SDGs to fulfill their corporate social responsibility goals.

The integration of the SDGs into the Olympics not only enhances the event's reputation but also signifies its commitment to addressing global challenges and making a positive impact on society. By embracing sustainable practices, promoting inclusivity, and supporting community development, the Olympics can further solidify its position as a catalyst for sustainable development and inspire positive change on a global scale.

Purpose of study and research questions

Sports organizations are perceived as taking environmental sustainability seriously and implementing meaningful sustainability initiatives. This can result in increased interest and support from fans and stakeholders. This study will address the social dimension of sustainability, in particular the "public support of the event," which is expressed as "public approval."

The hypothesis for this study is: *Positive sentiments regarding the SDGs are positively related to support for the Olympics. Negative sentiments to the SDGs are negatively related to support for the Olympics.*

[perception toward sustainability initiatives] → [support toward Olympics]

Methods

This study quantitatively examined the public sentiment toward the Beijing 2022 Winter Olympics through user-generated content on Twitter. Using a combination of topic modeling, sentiment analysis, and traditional statistical techniques, we answer the research question at hand from three different angles. First, by analyzing the salience of identified topics changing over time, we verify if the sustainability of the Winter Olympics is a matter of significance for the public, and if so, it receives sustained attention. Second, we examine the difference in overall sentiment score levels among topics to identify the public perception toward the Olympics' contribution to the SDGs (or lack thereof). Subsequently, trends in sentiment scores were analyzed to quantify whether the public sentiment on SDG-related topics improves or worsens with time. Lastly, we used entity-level sentiment analysis for a more fine-grained investigation. Specifically, we probed the public sentiment toward the Olympic games and host destination, the difference in the sentiment score level by topic, and the change in sentiment over time.

Data curation

English tweets related to Beijing 2022 Winter Olympics were collected using the keywords "2022 winter olympics" or "Beijing winter olympics." Initially, tweets posted between 14 November 2013 (the application deadline for XXIV Olympic Winter Games) and 31 May 2023 were retrieved using a custom scraping script, which resulted in 167,475 tweets. The number of tweets first peaked around 31 July 2015, when Beijing was announced as a host destination (see Figure 10.2).

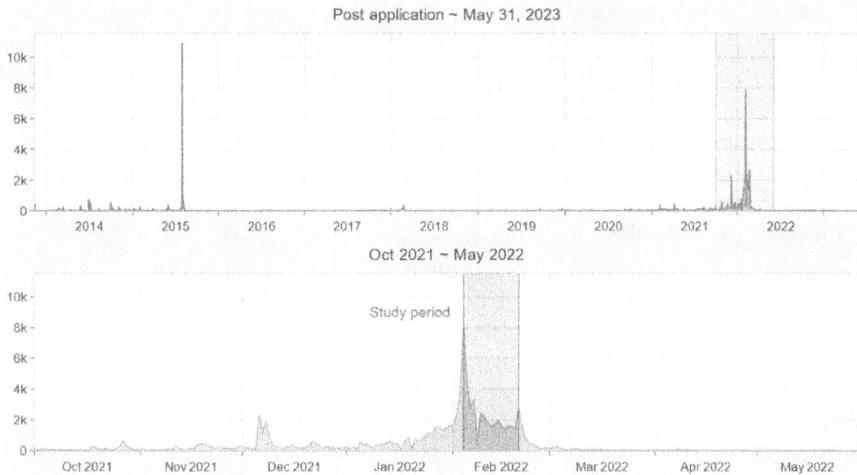

Figure 10.2 Evolution of Twitter activity about the Beijing 2022 Winter Olympics.

The next peak in Twitter activity occurred during the Beijing 2022 Winter Olympics period, between 4 and 20 February 2022. During these 17 days, users posted 40,767 tweets related to the Winter Olympics, taking approximately 24% of our full dataset over 10 years. Hence, the authors deemed that limiting the study period to the actual Olympics period can sufficiently capture the public attitude toward Beijing 2022 Winter Olympics. Emojis and regular expressions (e.g., URLs) were cleaned before conducting the analysis.

Topic modeling

Topic modeling is a process of identifying a common theme latent in documents (Kapoor et al., 2018). In this study, we used the BERTopic algorithm (Grootendorst, 2022) which utilizes the Bidirectional Encoder Representations from Transformers (BERT) language model. BERTopic algorithm has five main sequences for detecting the underlying topics. First, documents are transformed into embeddings, numerical representations of a given text. For our model, we used a pre-trained MPNet model (Song et al., 2020), which was trained with over 1 billion sentences. This process transformed the tweets dataset into a 768-dimensional vector space. Second, the dimension of embeddings is reduced to avoid the curse of dimensionality (Bellman, 1957) and for easier data handling. By default, BERTopic uses the Uniform Manifold Approximation and Projection (UMAP) technique (McInnes et al., 2020) for dimensionality reduction. Through this process, the dimensions of embeddings were reduced from 768 to 5. Third, the reduced embeddings are clustered into topic groups with similar characteristics. Again, we used the default Hierarchical Density-Based Spatial Clustering of Applications with Noise (HDBSCAN) model.

The last two steps create representations for the topics. In the fourth step, all documents within a cluster are concatenated into bag-of-words, which are then converted into a matrix of word frequency. Stop words (e.g., "a," "the," "of," and "is") were excluded for better interpretability of topic representations. Finally, the Term Frequency–Inverse Document Frequency (TF-IDF) statistic was calculated to choose important words that well-represent the topic. Words that frequently appear in every cluster (for instance, "Olympics" or "Winter" in our dataset) are removed to fine-tune the topic representation.

We chose to use the BERTopic algorithm for the following two advantages. BERTopic is highly modular, and different approaches can be employed interchangeably in each process. For instance, the PCA technique may be used instead of UMAP for dimensionality reduction, without affecting the subsequent steps. Another advantage is that the BERTopic algorithm does not force documents into topics (Grootendorst, 2022). Instead, documents that do not belong in any topics or clusters too small to be considered as topics are categorized as "outliers," improving the homogeneity of topics.

Sentiment analysis

This study employed sentiment analysis to measure the public support for the 2022 Winter Olympics. The automated sentiment analysis can be categorized into two types based on the approach: lexicon- and machine learning-based methods. The lexicon-based approach uses a set of affective words (i.e., sentiment lexicons) to detect emotions within a text. For example, in the sentence "I hate oranges," *hate* is an affective word associated with negative sentiment. Lexicon-based sentiment analysis algorithms differ not only in the coverage of lexicons but also in how the occurrence of affective words is weighted. This includes, but is not limited to, the use of negations, emojis, capitalization, punctuations, and slang (Kirilenko et al., 2018). Another approach to classifying sentiments in the text is to use a machine learning technique. This requires the model to be trained over pre-classified training data. One downside of this approach is that it is unclear how the sentiment scores were derived (Hur et al., 2022), unlike the lexicon-based approach which has explicit rules and lexicons. Different learning algorithms and training data can be used for machine learning-based approaches, which both have an impact on the performance of the model (Kirilenko et al., 2018).

Sentiment analysis can also be categorized on the basis of the scope of the analysis. Conventional sentiment analysis algorithms primarily focus on classifying the sentiment of the document. While this document-level analysis can help us quantify general attitudes within a text, the information it provides is limited because the target of the sentiment is unclear (Mitchell et al., 2013). This is especially relevant when there is more than one target within a text (e.g., "I hate oranges but I love apples!"). Entity-level sentiment analysis (also referred to as targeted sentiment analysis) aims to mitigate this issue by limiting the scope of the sentiment (Li & Lu, 2017). For instance, to use the previous example, negative sentiment *hate* is targeted toward *oranges*, but not *apples*. Thus, the sentiment scope of *hate* is limited to *oranges*.

For our analysis, we took a machine learning approach using a pre-trained model from Google Cloud Natural Language API (*Cloud Natural Language*, n.d.). Because this study will analyze sentiments toward the 2022 Winter Olympics at both document and entity levels, it was crucial to ensure consistency between the two levels of sentiment analysis. To the best of our knowledge, Google Cloud Natural Language is the only service that provides sentiment analysis at both levels.

Trend analysis

Testing the statistical significance of trends in time series data is commonly performed using Pearson correlation, Spearman rank correlation, or Mann–Kendall test (Tan & Gan, 2015). Although Spearman and Mann–Kendall tests are non-parametric and thus make no assumptions regarding the distribution of the data, the statistical power of Pearson correlation is stronger if the data are normally distributed (Gauthier, 2001). In this study, Pearson correlation was used to test the presence and directionality of trends in the data. Since the proportion of topics and sentiment scores are bounded continuous variables (ranging between $|\,0 \leq 1\,|$), we first unbounded these dependent variables using logit transformation. Because the range of sentiment scores is $[-1, 1]$, we first transformed the sentiment score x into $\dfrac{x+1}{2}$ before the logit transformation. The transformed variables met the normality criteria of skewness between -2 and 2 and kurtosis between -7 and 7. Although there was no notable difference caused by the transformation, the analysis result using the original values was also included for clarity.

Findings

Topic modeling

The trained topic model extracted 44 topics from 40,767 tweets related to the 2022 Winter Olympics. About 37% of tweets did not belong to any of the identified topics and were categorized as *Outliers*. The number of tweets in each topic and their representative words are summarized in Table 10.1. Researchers manually labeled topics based on word representations generated by the model. Due to the sparsity of the data points, topics 7–44 were inadequate for trend analysis; hence, they were merged and labeled as *Other*.

Of the 44 resulting topics, we identified that the following four clusters are related to SDGs. Topic 1 (*Human rights*, N = 2,135) is the cluster of tweets that discuss the accusation that China violated human rights against the ethnic minority group Uyghurs, which concerns *Goal 16 – Peace, justice, and strong institutions*. Increased political tension in the Asia region related to Russia–Ukraine relations and international cooperation through the Olympic Games are discussed in Topics 2 (*Russo-Ukraine*, N = 1,460) and 6 (*Unity*, N = 1,178), respectively. Tweets that

Table 10.1 Summary of trained topic model

Topic	Label	Representative wordsa	N	%
0	*Good luck*	Luck, wish, athletes, good . . .	2,327	5.7
1	*Human rights*	Boycott, human, rights, genocide . . .	2,135	5.2
2	*Russo-Ukraine*	Putin, Ukraine, president, xi . . .	1,460	3.6
3	*Hockey*	Hockey, Canada, Finland, vs . . .	1,607	3.9
4	*Climate change*	Snow, artificial, climate, fake . . .	1,533	3.8
5	*Opening ceremony*	Opening, ceremony, ceremonies, stadium . . .	1,373	3.4
6	*Unity*	Sports, unity, organization, people . . .	1,178	2.9
7	*Other*	Valieva, Kamila, doping, Russian . . .	1,161	2.8
8		Curling, gb, Britain, Sweden . . .	987	2.4
9		Stream, peacock, coverage, watch . . .	1,002	2.5
10		COVID-19, cases, positive . . .	922	2.3
11		DWEN, mascot, bing, panda . . .	647	1.6
12		Dance, figure, skating, pairs . . .	550	1.3
13		Eileen, gu, freeski, freestyle . . .	506	1.2
14		Shiffrin, Mikaela, slalom, Shiffrins . . .	602	1.5
15		Arif, Khan, Indian, Kashmir . . .	434	1.1
16		Closing, close, closes, Sunday . . .	492	1.2
17		Ratings, viewers, viewership, million . . .	421	1.0
18		Technology, robots, tech, robot . . .	430	1.1
19		Count, table, tally, medal . . .	395	1.0
20.44		. . .	5,486	13.5
Outliers	*Outliers*	–	15,119	37.1

Note: a Due to limited space, only the first four are shown here.

belong to these two topics regard *Goal 17 – Partnerships for the goals.* Lastly, Topic 4 (*Climate change*, N = 1,533) is a cluster of tweets that mentioned the use of artificial snow during the 2022 Winter Olympics and the impact of the event on climate, which is associated with *Goal 13 – Climate action*. Examples of tweets that belong to these four SDG-related topics are shown in Figure 10.3. All SDG-related topics are included in the top 10 in terms of the size of a cluster. When combined, these topics consist of about 15% of the entire tweets (25% if outliers are excluded). The high salience of SDG-related themes signifies the public interest in the sustainability of the Winter Olympics.

Trends in topic salience

After establishing the presence of SDG-related themes, we first examined how public attention on topics related to the Winter Olympics changes over time. The salience of each topic was calculated using the proportion from the daily number of tweets (Figure 10.4). Subsequently, trend analysis was performed using the proportion of topics. The "waning" of topics was visually identifiable, where the

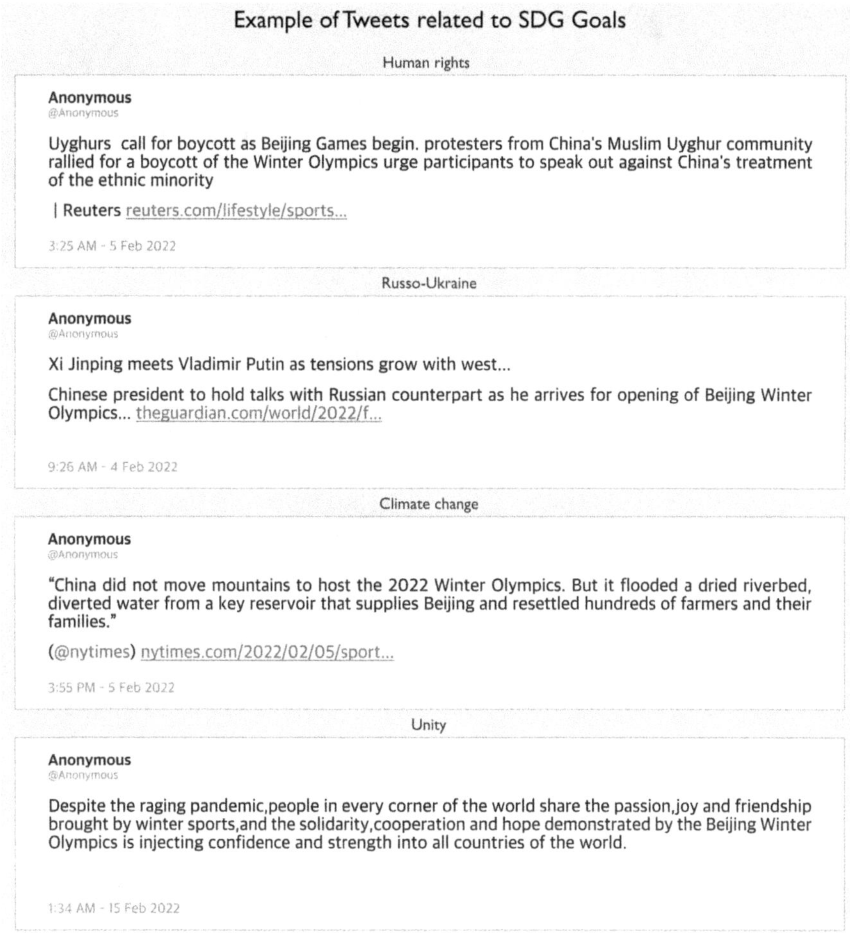

Example of Tweets related to SDG Goals

Human rights

Anonymous
@Anonymous

Uyghurs call for boycott as Beijing Games begin. protesters from China's Muslim Uyghur community rallied for a boycott of the Winter Olympics urge participants to speak out against China's treatment of the ethnic minority

| Reuters reuters.com/lifestyle/sports...

3:25 AM - 5 Feb 2022

Russo-Ukraine

Anonymous
@Anonymous

Xi Jinping meets Vladimir Putin as tensions grow with west...

Chinese president to hold talks with Russian counterpart as he arrives for opening of Beijing Winter Olympics... theguardian.com/world/2022/f...

9:26 AM - 4 Feb 2022

Climate change

Anonymous
@Anonymous

"China did not move mountains to host the 2022 Winter Olympics. But it flooded a dried riverbed, diverted water from a key reservoir that supplies Beijing and resettled hundreds of farmers and their families."

(@nytimes) nytimes.com/2022/02/05/sport...

3:55 PM - 5 Feb 2022

Unity

Anonymous
@Anonymous

Despite the raging pandemic,people in every corner of the world share the passion,joy and friendship brought by winter sports,and the solidarity,cooperation and hope demonstrated by the Beijing Winter Olympics is injecting confidence and strength into all countries of the world.

1:34 AM - 15 Feb 2022

Figure 10.3 Examples of tweets related to SDGs.

area of labeled topics almost halved at the end of the Winter Olympics compared to the beginning. The increasing salience of the *Other* category ($r = .698$, $p = .002$) and the negative correlation for four of seven remaining topics (Table 10.2) also indicate that the salience of topics generally decreases over time. The proportion of the *Human rights* topic showed the strongest negative correlation with time ($r = -.758$, $p < .001$), followed by *Opening ceremony* ($r = -.756$, $p < .001$) and *Russo-Ukraine* ($r = -.549$, $p = .022$). In contrast, *Climate change* showed the least correlation ($r = .011$, $p = .967$), followed by *Unity* ($r = .073$, $p = .782$), which indicates that the public's attention on these issues remained consistent throughout the Olympics period.

Cumulative Percentage of Tweets by Topic

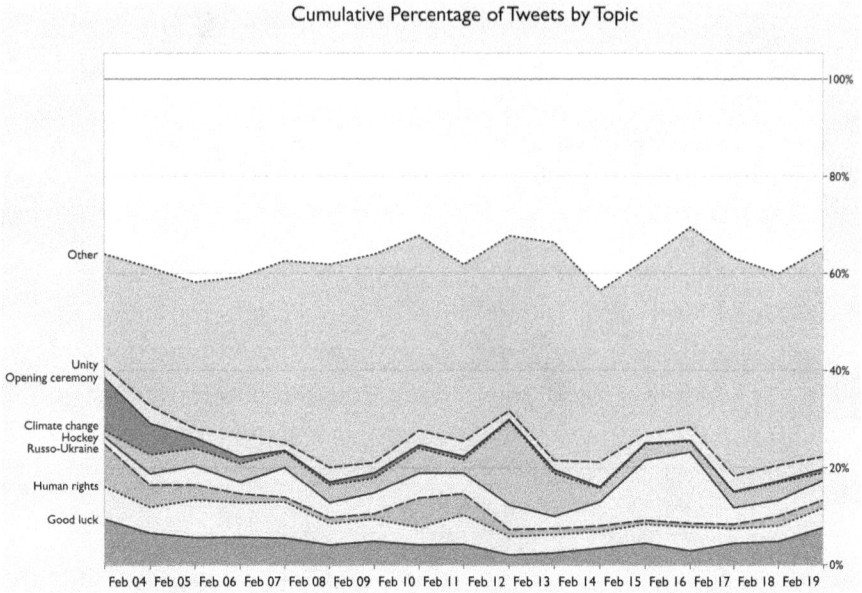

Figure 10.4 Percentage of tweets by date and topic.

Table 10.2 Trend analysis of topic proportion

Topic	Logit		Original	
Good luck	−.387		−.405	
Human rights	−.758	***	−.747	***
Russo-Ukraine	−.549	*	−.546	*
Hockey	.465		.397	
Climate change	.011		.037	
Opening ceremony	−.756	***	−.632	**
Unity	.073		.054	
Other	.698	**	.695	**

Note: $* p < .05, ** p < .01, *** p < .001.$

Trends in general sentiment in tweets

One-sample *t*-test result revealed that all eight topics showed statistically significant differences in mean sentiment scores compared with the baseline of average sentiment in all 40,767 tweets (Table 10.3). The topic *Human rights* showed the greatest difference in document-level sentiment scores ($t = -53.896, p < .001$), followed by *Climate change* ($t = -18.843, p < .001$) and *Russo-Ukraine* ($t = -14.722, p < .001$). In other words, the public perceived that the 2022 Winter Olympics has

Table 10.3 Analysis of mean document-level sentiment

Topic	Mean	SD	μ	MD	t	df	p
				One-sample t -test			
Good luck	.243	.314	.072	.171	26.189	2326	<.001
Human rights	−.284	.306		−.357	−53.896	2134	<.001
Russo-Ukraine	−.026	.255		−.098	−14.722	1459	<.001
Hockey	.147	.260		.075	11.511	1606	<.001
Climate change	−.082	.320		−.154	−18.843	1532	<.001
Opening ceremony	.183	.292		.110	13.999	1372	<.001
Unity	.240	.306		.168	18.863	1177	<.001
Other	.060	.303		−.012	−4.802	14034	<.001

Note: SD = standard deviation, MD = mean difference.

failed to achieve the SDGs in these aspects. Among SDG-related themes, *Unity* was the only topic that showed higher sentiment scores compared to the total average ($t = 18.863, p < .001$). This difference in sentiment score levels was also visible in the daily time series dataset (see Figure 10.5).

Tweets data were converted into daily time series using mean sentiment values for the trend analysis. The analysis result indicates that while the negative sentiment scores for the *Other* category improved over time, negativity in SDG-related topics (i.e., *Human rights, Russo-Ukraine,* and *Climate change*) did not wane with time (Table 10.4). In the case of *Russo-Ukraine*, the sentiment scores even showed a declining trend ($r = −.510, p = .036$). Nonetheless, positive sentiments in tweets are also static, as topics that showed above-average sentiment levels (including the SDG-related topic *Unity*) showed no statistically significant correlation with time.

Trends in sentiment toward the Olympics and host destination

Lastly, we examine the public attitude toward the Olympics and the host destination using entity-level sentiment scores. Similar to document-level analysis, sentiment scores toward the Olympics and the host destination in *Human rights, Russo-Ukraine,* and *Climate change* topics sit below the mean level, whereas sentiment levels for *Unity* are above the mean (see Table 10.5 and Figure 10.6). The difference in sentiment scores between SDG-related topics and the baseline is more prominent in Olympics-related entities, with greater t-values. This can be interpreted that the positive and negative sentiments of the public regarding the sustainability of the Winter Olympics are more strongly expressed toward the Olympic Games and IOC than the host destination.

Again, negative sentiments toward the Olympics and the host destination in the *Other* category improved with time, whereas negative scores in the SDG-related

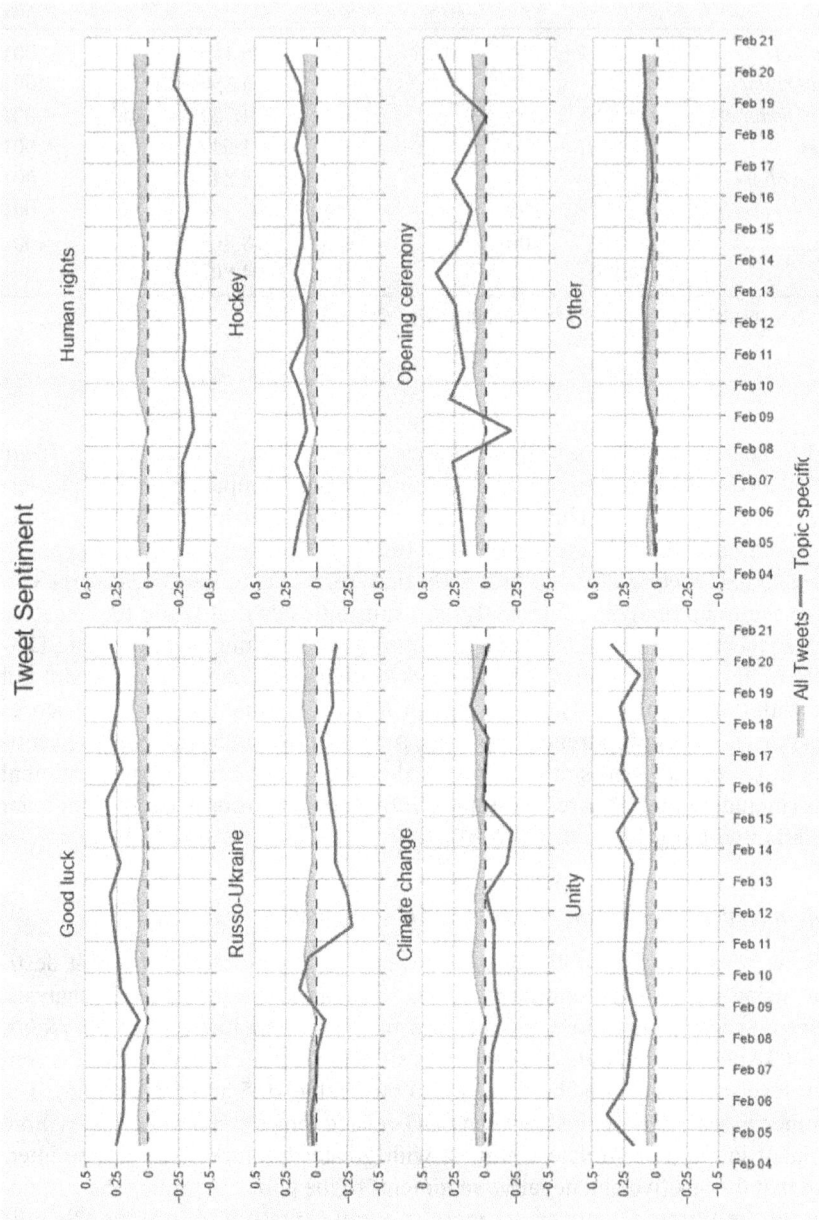

Figure 10.5 Document-level sentiment of tweets by date and topic.

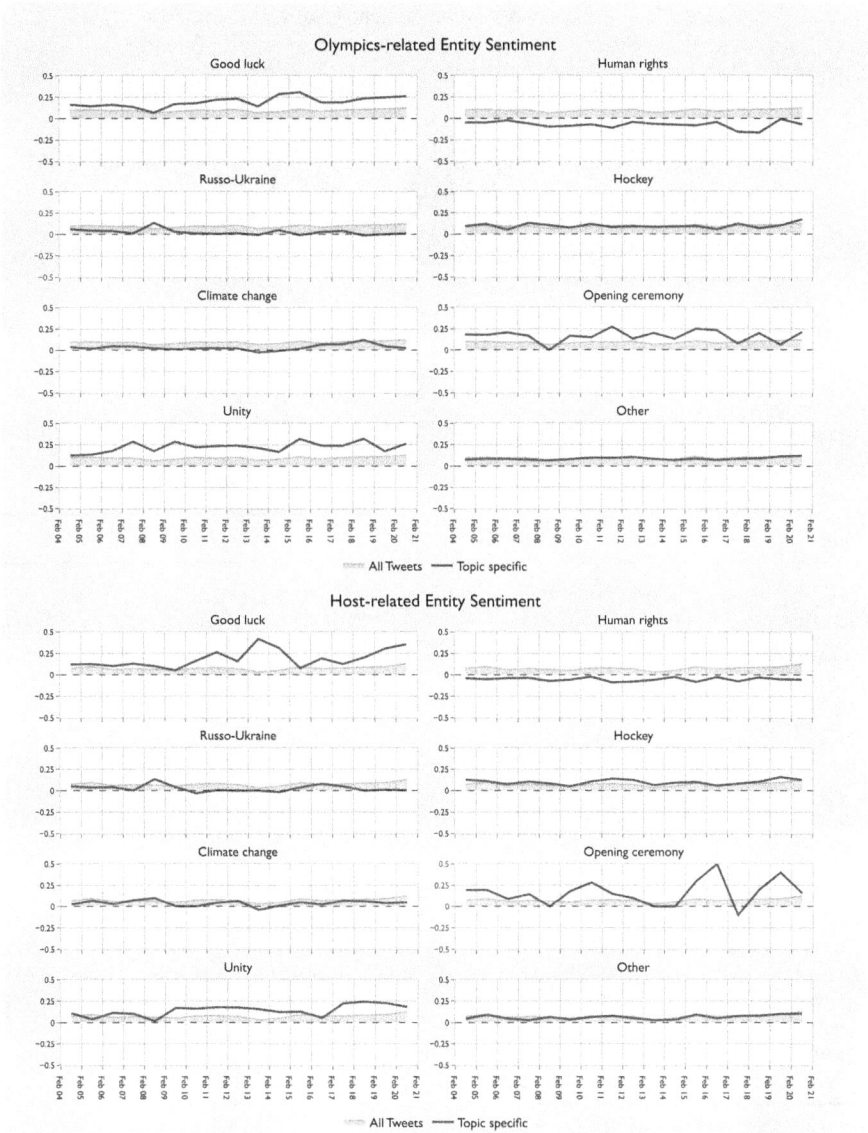

Figure 10.6 Sentiment of Olympics- and host-related entities in tweets by date and topic.

topics *Human rights*, *Russo-Ukraine*, and *Climate change* showed no significant correlation with time (Table 10.6). Combining the document- and entity-level analysis results, we can deduce that the perception of the Olympic Games failing to contribute to the achievement of SDGs leads not only to the public's expression of negative sentiments in general but also to negative sentiments toward the Olympics itself and the host destination. Moreover, this negative sentiment does not dwindle with time, unlike negativity in other topics.

Table 10.4 Trend analysis of document-level sentiment

Topic	Logit		Original	
Good luck	.478		.476	
Human rights	.138		.143	
Russo-Ukraine	−.510	*	−.516	*
Hockey	.174		.172	
Climate change	.364		.367	
Opening ceremony	.166		.159	
Unity	.099		.100	
Other	.571	*	.572	*

Note: * $p < .05$, ** $p < .01$, *** $p < .001$.

Table 10.5 Analysis of mean entity-level sentiment

Topic	Mean	SD	One-sample t-test				
			μ	MD	t	df	p
	Olympics-related entities						
Good luck	.182	.226	.098	.084	17.452	2202	<.001
Human rights	−.061	.194		−.159	−36.856	2017	<.001
Russo-Ukraine	.042	.106		−.056	−19.646	1373	<.001
Hockey	.099	.148		.001	0.196	1524	.845
Climate change	.019	.152		−.079	−19.686	1451	<.001
Opening ceremony	.183	.202		.085	14.484	1170	<.001
Unity	.213	.229		.115	16.811	1112	<.001
Other	.088	.161		−.010	−7.429	13193	<.001
	Host-related entities						
Good luck	.155	.207	.077	.078	8.772	538	<.001
Human rights	−.047	.165		−.124	−23.514	969	<.001
Russo-Ukraine	.039	.097		−038	−11.716	877	<.001
Hockey	.104	.147		.027	3.399	345	<.001
Climate change	.025	.154		−.053	−9.922	843	<.001
Opening ceremony	.190	.215		.113	11.704	496	<.001
Unity	.119	.196		.042	4.553	450	<.001
Other	.060	.152		−.017	−6.819	3617	<.001

Note: SD = standard deviation, MD = mean difference.

One notable difference in the entity-level analysis is that positivity toward the host destination in the topic *Unity* exhibited an increasing trend ($r = .617, p = .008$). That is, for the aspect of the Winter Olympics that the public perceived as a contribution to SDGs, people expressed increasingly positive emotions regarding the host destination. A similar trend can also be found in the entity-level sentiment score toward the Olympics, although it was not statistically significant ($r = .462, p = .062$).

Table 10.6 Trend analysis of entity-level sentiment

Topics	Olympics-related				Host-related			
	Logit		Original		Logit		Original	
Good luck	.655	**	.656	**	.533	*	.539	*
Human rights	−.279		−.278		−.073		−.073	
Russo-Ukraine	−.466		−.467		−.239		−.239	
Hockey	.174		.172		.149		.148	
Climate change	.278		.277		−.024		−.024	
Opening ceremony	.005		.003		.195		.179	
Unity	.462		.465		.617	**	.614	**
Other	.490	*	.490	*	.491	*	.491	*

Note: * $p < .05$, ** $p < .01$, *** $p < .001$.

Discussion

The findings of this research study offer crucial insights into the public perception of the SDGs in relation to the 2022 Winter Olympics. A critical observation was the substantial public interest in the SDGs related to the event. This was evident in the salience of SDG-related topics in the Twitter data, which included Human rights, Russo-Ukraine relations, Unity, and Climate change. These topics comprised a significant proportion of the overall tweets, underscoring their importance to the public discourse surrounding the Winter Olympics.

However, as the Winter Olympics unfolded, there was a general decrease in public attention to most of the topics. This "waning" was particularly marked for topics such as Human rights and Russo-Ukraine relations, pointing to a diminishing public engagement with these issues as the event progressed. Practically, this suggests that stakeholders should consider strategies to maintain public interest and engagement in these crucial topics throughout the course of such international events.

On a positive note, certain SDGs, specifically Climate change and Unity, maintained a relatively consistent level of interest throughout the event. The consistent public interest in these issues suggests they remain forefront in the minds of the public, even as other topics see diminished attention. Consequently, these areas may warrant particular attention from policymakers, organizers, and other stakeholders, not only during the event but also in its aftermath and planning for future events.

One of the more concerning insights derived from the sentiment analysis was the prevailing perception that the 2022 Winter Olympics fell short in addressing several key SDGs. Particularly in the areas of human rights, climate change, and Russo-Ukraine relations, there was a dominant negative sentiment both in general tweets and in those specifically related to the Olympics and the host destination. Unlike diminishing negative sentiments in other topics, the public's negative perception toward failures in these SDG-related areas did not improve with time. This perception of failure has implications for future events, indicating that organizers should take more transparent and tangible steps to address these SDGs to improve public sentiment.

Despite the predominantly negative sentiments, the study revealed that positive sentiments remained relatively static, showing no significant correlation with time. The perceived contribution to the SDGs in some areas, such as the topic of Unity, generated consistent positive sentiment. This emphasizes the value of pursuing and clearly communicating the achievements of these goals during such international events.

Another significant observation was the increasing positive sentiment toward the host destination within the Unity topic. While negative sentiments were generally more prevalent, this growing positive perception suggests that the public appreciates and rewards perceived contributions to SDGs, providing increasingly positive feedback about the host destination when they observe a commitment to these goals.

However, it is important to note that negative sentiments toward the Olympics and the IOC were more strongly expressed when the event was perceived as failing to contribute to SDGs. This finding suggests that both the games and the IOC need to take a proactive approach to integrate and demonstrate commitment to SDGs in order to enhance public sentiment.

In sum, this study's findings illustrate the crucial role that public perception plays in evaluating the success of major events in addressing key social and environmental issues. They underscore the need for tangible, demonstrable action on SDGs and for maintaining transparency and public engagement throughout such events. These insights can inform future planning and policies for similar international events to ensure they are more aligned with SDGs and, in turn, more positively received by the public.

Conclusion

The extensive analysis and discussion conducted in this study have provided us with a more profound understanding of the complexities and nuances of public sentiments regarding the Olympic Games in the context of SDGs. When the study was embarked upon, we explored the potential of the Olympics, a global mega event, to shape perceptions on sustainable development and the progress toward the attainment of SDGs. Given the strategic emphasis placed by the IOC on sustainability and the identification of 11 pertinent SDGs for the Olympic Movement, there was an implicit expectation that the games should contribute meaningfully toward the global sustainable agenda.

We noted that positive public sentiments toward the Olympics could be driven by perceptions of social progress, such as contributions toward reducing poverty, enhancing educational accessibility, promoting gender equality, and improving health outcomes. For the Olympic Games to resonate positively with the public, the games should be seen to promote not only sports but also an array of social goods. The notion of the Olympic Games contributing toward a better world was highlighted in the Olympic Agenda 2020 strategy, which emphasized the principles of sustainability. Consequently, alignment of the games with SDGs could engender a sense of global cooperation and purpose, leading to favorable public sentiments.

The study's findings, however, present a somewhat different narrative. While there was substantial public interest in the SDGs, especially those related to Human Rights, Russo-Ukraine relations, Unity, and Climate Change, sentiments associated with these topics were not universally positive. Indeed, the perception emerged that the 2022 Winter Olympics had not adequately addressed several of these key SDGs, leading to a dominance of negative sentiments in the related tweets. These sentiments were particularly strong toward the Olympics and the IOC, highlighting a perceived gap between the Olympic Movement's sustainable agenda and the actual realization of these goals.

However, amidst these challenges, we also found evidence of enduring positive sentiments. Topics like Climate Change and Unity, which retained consistent public interest, were indicative of the public's sustained commitment to these global issues. Furthermore, the topic of Unity showed consistent positive sentiment scores, underscoring that the public did perceive certain aspects of the Olympic Games as contributing positively to SDGs. An interesting and promising trend was the growing positive sentiment toward the host destination within the Unity topic, suggesting that tangible contributions to SDGs were appreciated and rewarded with increasingly positive feedback.

In retrospect, while the Olympic Games have taken strides toward integrating sustainable practices and addressing climate change through carbon offsetting, renewable energy integration, sustainable infrastructure development, and raising public awareness, our study suggests that more needs to be done. The public is clearly engaged and invested in the sustainable outcomes of these events, and their perception is crucial in evaluating the success of the games. It is evident that where the Olympic Games are perceived to fall short, negative sentiments arise, sometimes persistently. Conversely, where they are seen to contribute positively toward the SDGs, they are met with approval and even growing positivity.

In conclusion, the study illuminates the critical role of public sentiment and perception in the Olympic Games' sustainable agenda. It underscores that the real and perceived contributions of the Olympic Games toward the attainment of SDGs have real-world implications on the public's perception and reception of the event. It highlights the need for more than just nominal commitment to these goals but a proactive, tangible, and transparent approach to integrating and demonstrating commitment to SDGs. As the Olympic Games continue to evolve, the lessons from this study can contribute toward making future games more sustainable, more attuned to the global SDGs, and, consequently, more positively received by an increasingly discerning and engaged public.

Appendix 1

Goal 1: End poverty

According to the United Nations, poverty rates went down by half since the year 1990. Despite this achievement, one in five people reportedly still live on $1.25 or less daily. The goal now is to exterminate poverty entirely.

Goal 2: Zero hunger

One in nine people around the world is malnourished. More shockingly, a majority of this population live in developing countries. Ending global hunger is not the only goal noted. Ending malnutrition and promoting sustainable agriculture around the world are a part of this.

Goal 3: Good health and well-being

This goal primarily deals with the improvement of people's health regardless of age and gender. Improving modern medication and reducing the spread of HIV/AIDS, malaria, polio, and tuberculosis are also a big part of this goal. This goal also aims to reduce maternal mortality rates and end preventable deaths of children below 5 years of age and newborns.

Goal 4: Quality education

The target here is to ensure all students, regardless of gender, complete free primary and secondary education. Access to education has always been a problem. The UN wants to solve that. The quality of said education is also important. Improving the quality of free education by the year 2030 is one of the biggest aspects of this goal.

Goal 5: Gender equality

This goal has two main sides to it. The first side is to ensure all women get equal access to quality health care, work, education, and positions of power in the government and society. The other side is to ensure all women are protected from violence, sexual exploitation, and other forms of discrimination. Goal 5 is also integrated into all of the other SDGs.

Goal 6: Clean water and sanitation

Reducing water pollution and ensuring access to clean and affordable drinking water are both main targets. By achieving Goal 6, the other goals will be attained. Proper water sanitation and distribution will aid in reducing diseases and improving the health of the population.

Goal 7: Affordable and clean energy

Doubling the global rate of energy efficiency improvement is one of the main goals listed by the UN. Other goals include increasing renewable energy sources, developing cleaner and safer fossil fuels, and ensuring universal access to affordable and reliable energy sources.

Goal 8: Decent work and economic growth

By 2020, the UN aims to reduce the amount of unemployment in the youth. By 2030, the UN aims to provide equal pay for men and women. Also, by 2030, the target is to provide equal work opportunities and to eliminate forced labor.

Goal 9: Industry, innovation, and infrastructure

This goal primarily focuses on the improvement of industrial development. This includes better communication technology, information technology, energy, and transport.

Goal 10: Reduced inequalities

Reducing inequalities connects back to Goal 1. Ending poverty does not simply mean improving the economic growth. To reduce income inequalities, the UN states each country must improve social and environmental development as well.

Goal 11: Sustainable cities and communities

The main target of this goal is to make cities safe to live in. Each city around the world must also be resilient; inclusive to people of all ages, gender, and cultural background; and must be economically sustainable.

Goal 12: Responsible consumption and production

According to the UN, the target here is to "do more and better with less." We can all do better by 2030 by reducing pollution and resource use. Promoting environment-friendly and decent jobs also fall under this goal.

Goal 13: Climate action

Climate change is a hotly debated issue around the world. The UN aims to fight climate change and its impacts on the world. In addition to focusing on renewable energy sources, each nation also signed and adopted the Paris Agreement. This document states each member of the UN will work to limit global temperature rise to below 2 degrees Celsius.

Goal 14: Life below water

Water pollution is the main cause of disruption of marine life. The UN aims to significantly reduce water pollution by 2030. Another main target is to improve aquaculture, tourism, and the use of marine resources.

Goal 15: Life on land

Halting biodiversity loss and reversing land degradation are key components of this goal. Countries also aim to sustain manageable forests, reduce land pollution, fight desertification, and promote improved use of terrestrial resources.

Goal 16: Peace, justice, and strong institutions

This goal aims to reduce all forms of crime, including all forms of violence, abuse, corruption, and bribery. It also aims everyone can avail of equal justice and the rule of law.

Goal 17: Partnerships for the goals

None of these goals can be achieved if the members of the UN do not work together. This goal aims to promote the partnership of each member. This includes providing monetary assistance to developing countries and providing better trade between each member.

References

Agarwal, A., Xie, B., Vovsha, I., Rambow, O., & Passonneau, R. J. (2011, June). *Sentiment analysis of Twitter data*. Proceedings of the workshop on language in social media (LSM 2011), 30–38.

Beek, R., Van Hoecke, J., & Derom, I. (2023). Sponsorship and social justice: Brand positioning on diversity and inclusion in sport marketing during the 2020 UEFA European Football Championship. *International Journal of Sports Marketing and Sponsorship*, 24(3), 538–557. https://doi.org/10.1108/IJSMS-03-2022-0069

Bellman, R. (1957). Dynamic programming. Princeton, New Jersey: Princeton University Press.

Cloud Natural Language. (n.d.). *Google Cloud*. Retrieved July 4, 2023 from https://cloud.google.com/natural-language

Gasser, P. K., & Levinsen, A. (2004). Breaking post-war ice: Open fun football schools in Bosnia and Herzegovina. *Sport in society*, 7(3), 457–472.

Gauthier, T. D. (2001). Detecting trends using Spearman's rank correlation coefficient. *Environmental Forensics*, 2(4), 359–362. https://doi.org/10.1006/enfo.2001.0061

Grootendorst, M. (2022). BERTopic: Neural topic modeling with a class-based TF-IDF procedure. *arXiv*, 2203.05794. https://doi.org/10.48550/arXiv.2203.05794

Hur, D., Lee, S., & Kim, H. (2022). Are we ready for MICE 5.0? An investigation of technology use in the MICE industry using social media big data. *Tourism Management Perspectives*, 43, 100991. https://doi.org/10.1016/j.tmp.2022.100991

International Olympic Committee (IOC). (n.d.). *Olympic Agenda 2020*. Retrieved from www.olympic.org/olympic-agenda-2020

International Olympic Committee. (2012). *London 2012 Olympics: Achieving carbon neutral games*. Retrieved from www.olympic.org/london-2012/achieving-carbon-neutral-games

Kahneman, D., Fredrickson, B. L., Schreiber, C. A., & Redelmeier, D. A. (1993). When more pain is preferred to less: Adding a better end. *Psychological Science*, 4(6), 401–405.

Kapoor, K. K., Tamilmani, K., Rana, N. P., Patil, P., Dwivedi, Y. K., & Nerur, S. (2018). Advances in social media research: Past, present and future. *Information Systems Frontiers*, 20(3), 531–558. https://doi.org/10.1007/s10796-017-9810-y

Kirilenko, A. P., Stepchenkova, S. O., Kim, H., & Li, X. (Robert). (2018). Automated sentiment analysis in tourism: Comparison of approaches. *Journal of Travel Research*, 57(8), 1012–1025. https://doi.org/10.1177/0047287517729757

Li, H., & Lu, W. (2017). *Learning latent sentiment scopes for entity-level sentiment analysis*. Proceedings of the AAAI Conference on Artificial Intelligence, 31(1), Article 1. https://doi.org/10.1609/aaai.v31i1.11016

Manzenreiter, W., & Horne, J. (2005). Public policy, sports investments and regional development initiatives in Japan. *The political economy of sport*, 152–182.

McInnes, L., Healy, J., & Melville, J. (2020). UMAP: Uniform manifold approximation and projection for dimension reduction. *arXiv*, 1802.03426. https://doi.org/10.48550/arXiv.1802.03426

Mitchell, M., Aguilar, J., Wilson, T., & Van Durme, B. (2013). *Open domain targeted sentiment*. Proceedings of the 2013 Conference on Empirical Methods in Natural Language Processing, 1643–1654. Retrieved from https://aclanthology.org/D13-1171

Müller, M., Wolfe, S. D., Gaffney, C., Gogishvili, D., Hug, M., & Leick, A. (2021). An evaluation of the sustainability of the Olympic Games. *Nature Sustainability*, 4(4), 340–348.

Reyes-Menendez, A., Saura, J. R., & Alvarez-Alonso, C. (2018). Understanding #WorldEnvironmentDay user opinions in Twitter: A topic-based sentiment analysis approach. *International Journal of Environmental Research and Public Health*, 15(11), 2537.

Schulenkorf, N. (2012). Sustainable community development through sport and events: A conceptual framework for sport-for-development projects. *Sport Management Review*, 15(1), 1–12.

Song, K., Tan, X., Qin, T., Lu, J., & Liu, T.-Y. (2020). MPNet: Masked and permuted pretraining for language understanding. *arXiv*, 2004.09297. http://arxiv.org/abs/2004.09297

Tan, X., & Gan, T. Y. (2015). Nonstationary analysis of annual maximum streamflow of Canada. *Journal of Climate*, 28(5), 1788–1805. https://doi.org/10.1175/JCLI-D-14-00538.1

Thelwall, M., Buckley, K., & Paltoglou, G. (2012). Sentiment strength detection for the social web. *Journal of the American Society for Information Science and Technology*, 63(1), 163–173.

Tokyo 2020. (2021). *Sustainable sourcing code*. Retrieved from https://tokyo2020.org/en/games/sustainability/sustainable-sourcing-code/

Tumasjan, A., Sprenger, T., Sandner, P., & Welpe, I. (2010, May). *Predicting elections with Twitter: What 140 characters reveal about political sentiment*. Proceedings of the International AAAI Conference on Web and Social Media, 4(1), 178–185.

UNEP. (2020). *Sports and sustainability: A guide to greener sporting events*. Retrieved from www.unep.org/resources/report/sports-and-sustainability-guide-greener-sporting-events

Wylleman, P. (2019a). A developmental and holistic perspective on transitioning out of elite sport. In M. H. Anshel (Ed.), *APA handbook of sport and exercise psychology: Vol. 1. Sport psychology* (pp. 201–216). American Psychological Association.

11 Fostering peaceful and respectful societies with sports tourism and events

Miroslav Knezevic

Introduction

Sports and tourism, from the first sporting events (the organization of the first Olympic Games in Olympia, the Pythian Games in Delphi, the Isthmian Games in Corinth, and the Nemean Games in Nemea) until today, are interconnected, and one can hardly be observed without the other. Sports have become an increasingly popular activity, which is why more and more people travel who want to participate in sports activities, competitions, matches, championships, and other events in any way. In modern tourism, sports become not only the content of the stay but often also the main motive for traveling to certain tourist destinations (Bartoluci, 2004). Considering that there are various national and world cups, championships, competitions, and sports events that are held in different parts of the world, in order for athletes to be able to participate in sports events, they must travel and stay in a certain country (city) for a certain number of days, more precisely until their participation in the sporting event is over. Also, spectators, fans, coaches with professional staff, doctors, journalists, and all others who participate in sports in any way, and are not athletes, must also travel to the place where the sports event takes place and stay there for several days. In this way, they are considered tourists because they leave their place of permanent residence and go to another place or country where they stay for a few days, meet the local population, experience gastronomic specialties, spend the money they have earned in the place they come from, and thus create multiplied impacts to society and the place where they live. Therefore, if there was no sports, there would not be so much travel, and on the other hand, if there were no tourist infrastructure and the opportunities provided by tourism, sports events would not be able to take place in various parts of the world, and athletes and other personnel would not be able to stay in them. For this reason, it is considered that sports and tourism are dependent on each other, that they are mutually complementary, and that they can significantly influence the local population as well as local economic development. Sports in modern tourism do not only have an observational role, but they, with all their contents and forms, have become an important content of tourists' stay in which they become active participants in various sports and recreational activities (Mato & Nevenka, 1998). Nowadays, the tourism industry records various economic, social, political, and other influences

DOI: 10.4324/9781003384786-11

that provide it with an important position in the overall economic development (Čerović et al., 2015) if we include sports events and the organization of various events that gather a large number of people. We can conclude that the impacts and contribution of these activities to the development of society are extremely significant. Immense attention, in numerous studies (Čerović et al., 2016), has been devoted to the analysis of the economic effects of tourism and other mentioned activities on destinations; however, other impacts on society must not be neglected either, primarily better understanding between people, the spread of peace, and building harmonious relationships that affect overall social development (Zhuang et al., 2019). Ever since 1966, when the UN proposed a resolution and declared 1967 as the International Year of Tourism, tourism has been promoted as a "passport to peace" (Pedersen, 2020). Today, more than 55 years since then, tourism still plays that role – to reconcile and to help people get to know other destinations, cultures, peoples, customs, and other people who live there better. Thanks to this, tourism, along with sports, is becoming recognized as an important element for achieving the Sustainable Development Goals (SDGs) defined by the United Nations Agenda (United Nations General Assembly, 2015). Accordingly, in this chapter, we will show how, thanks to sports, tourism, and various events (primarily sports), positive relations in society can be nurtured and developed, some conflicting situations can be overcome, and sports can contribute significantly to the inclusion of individuals and sensitive communities, as well as to better understanding, education, and the goals of social inclusion – which makes an immeasurable contribution to the achievement of the aforementioned goals of sustainable development.

Sports, tourism, and events – definition, significance, and impacts on the social community

From a historical point of view, in the past, tourism was mostly available to the "higher" layers of the society, that is, the privileged classes, but with the advent of industrialization and modernization (Bianchi & Stephenson, 2013), that is, the creation of "excess" free time for people, and with their increased desire to meet, visit, see, and experience other, new destinations and ways of living, tourism today is no longer a luxury but has become a widespread phenomenon accessible to all segments of the society (MacCannell, 2013). Therefore, tourism as a socioeconomic phenomenon represents one of the basic functions of the expression of free time in modern society, and on the basis of this, it creates additional effects that promote overall economic and social development (Telfer & Sharpley, 2015). It is noticeable that the movements of "traveling humanity" have been influencing the formation of a specific tourist culture as a temporary way of life for years (Hussain, 2021). Today, tourism represents one of the most dominant social and economic phenomena of the modern era (Higgins-Desbiolles, 2006), which affirmed the knowledge of a different reality and way of life that people identify with (Butler, 2015). Tourist trips actually represent an attempt to minimize and even cancel alienated forms of existence that are imposed by everyday life. In theory and literature, tourism is most often seen as a human need (Cohen, 1979) arising from the modern way of life (Wang, 1999; Aho, 2001; Larsen, 2019). As a result of globalization and intense competition, all countries are re-examining their strategies and policies in order to raise the awareness of the global public, become attractive, and thus present themselves as orderly systems in which respect, peace, and stability reign, accordingly attracting investors, customers, tourists, and visitors. In accordance with the mentioned characteristics, according to the observations of Hollinshead and Hou (2012), tourism has been assigned a prominent role as a producer, communicator, representative, and promoter of a country. These authors emphasize the role of tourism and its contribution presented through the realistic perception of people (culture, customs, habits, and other characteristics) and places, and through everyday institutional and corporate interaction (social and personal). Accordingly, the authors propose to follow and develop studies according to which tourism articulates the space and lifestyle, develops a creative image of the population and places, and performatively depicts populations and places (Hollinshead & Xiao Hou, 2012).

In this context, connecting sports and tourism (Van Rheenen et al., 2017) represents a refreshing oasis for the so-called modern traveling humanity. Tourism, like sports, is characterized by a specific need to be active, to act, and to be innovative and different. Sports, and especially large sports events since ancient times (Unković & Zečević, 2016), have a positive effect (Higham, 1999), encouraging people to travel, providing them with active or passive participation in sports events (Čerović et al., 2022). Sports manifestations that provide not only interesting content but also the possibility of an active vacation to their visitors represent one of the main components of sports tourism (Gammon & Robinson, 2003), and perhaps the most significant in terms of the number of tourists and economic impact. As proof of such claims, a study of Western European countries was conducted in the eighties of the last century, which showed a strong growth of interest in the field of recreational sports. In addition to interest, the participation of all social groups also increased, and there was also an increase in both young participants and middle-aged participants during non-working days (holidays, school days off, etc.). Looking at the results of the study, it became clear that this phenomenon cannot be reduced to the usual interpretations as "foreigner traffic" and behavior in free time.

Today, the joint influence of sports, sports events, and tourism as an industry is increasingly pronounced (Mollah et al., 2021), which resulted in the emergence of sports tourism (Gozalova et al., 2014) which was conceived on the basis of analyzing sports as an important element of tourist trips (Hinch & Higham, 2001) and, at the same time, as a driver of the movement of a large number of travelers in the world. Sports tourism represents a very interesting and growing field (Jiménez-García et al., 2020) that has a unique ability to attract a large number of visitors, providing participants (travelers) who are looking for a sports experience to experience a sport that, due to its scope, complexity, and potential, has the potential to grow into a completely new industrial area (Borovcanin & Lesjak, 2022). In line with these claims are numerous studies based on data from the United Nations World Tourism Organization (UNWTO), according to which the field of sports tourism generated extremely significant revenues (Pedauga et al., 2022) and at the same time contributed to the sustainable development of tourist destinations (Morfoulaki et al., 2023). The processes of globalization and democratization have had a significant impact on the increase in consumption and development processes, as well as on sports tourism and tourism in general.

Several strategies are available to put a destination on a tourist's mental map, one of which is organizing events (Judd & Fainstein, 1999). Events, especially mega events, have often been argued to be image builders and play an important role in helping build a unique image of destinations that is different from competitors (Mossberg, 2000). Events play the role of animators, which contribute to making the destination seem warm and friendly, and as such are able to deliver key messages about the place and a positive image of the community to the world. Therefore, events are not only a way to attract tourists to a destination to spend money on accommodation, food, and activities at the destination but also a way to help build an image in the minds of tourists who have never visited a certain place. The intention and characteristics of the event are most often taken as the criteria for event classification. Accordingly, Getz distinguishes eight types of events (Getz, 2008):

- Cultural events
- Art/entertainment events
- Business/trade events
- Sports events
- Educational and scientific events
- Recreational events
- Political/state events
- Private events (anniversaries, celebrations, club parties, family gatherings, weddings, etc.).

Thus, events can manifest intangible heritage (customs, gastronomy, and marking of important dates); cover historical, sports, film, music, entertainment, artistic, scientific, political, cultural, or life topics; and bring them back to life for visitors and local residents. In this way, you get to know and better understand tourists and the local population, which can result in building a positive attitude about a destination and the population that lives in a certain area (Stylidis et al., 2022). Research by Erfurt and Johnsen (2003) showed that events provide a large number of tourists high exposure to the public through wide media coverage, and they especially pointed out that the organization of events has a significant impact on the traveler's attitude about the image of the destination. Without going into the classification, typology, and more detailed explanations of the events below, we will pay attention to sports events that, in addition to having great economic importance for the development of certain destinations, have a significant impact on the positive development of a society as a whole.

The category of sports events, which at the same time represent a tourist product, refers to sports activities that attract a significant number of visiting participants and spectators. The types of attendees vary by sporting event, and some are obviously more spectator-driven than others (e.g., the Olympics vs. the National Amateur Bowling Championships). High-profile sporting events such as the Super Bowl, the Olympics, or the World Cup have the potential to attract a large number of nonresidents, media and technical personnel, professional public such as coaches and other sports officials, and a large number of prominent individuals from the world of sports, business, politics, etc. Thanks to this and also to the large investments that precede it, primarily in the construction of sports infrastructure, certain sports events have a significant economic and then social and cultural impact (González-García et al., 2022), so the authors more and more often speak of them as an effective means of stimulating the overall development of the local environment (Zagnoli & Radicchi, 2009; Tsekouropoulos et al., 2022).

Social impacts of sports, sports tourism, and sports events

Historically, sports have played an important role in all societies and acted as a powerful communication platform that can be used to promote adequate social values and a culture of peace. Because of this, they are also often presented as one of the most cost-effective and versatile tools for promoting the values of the United Nations and achieving the SDGs.

The United Nations' SDGs stand out as a common language that unifies the global commitment to change the trajectory of social, economic, and environmental development issues. Although not the most directly cited within the SDGs or their related goals, sports are widely accepted and promoted as a driver of social change and a mechanism through which commitment to sustainability is strategically mapped and measured (Morgan et al., 2021). The vivid support and advocacy of global organizations, such as the UN, for sports as a means to address several goals contained in the SDGs have encouraged a large number of organizations (both private and public) to integrate sports as a cultural tool to contribute to their achievement (Collison et al, 2017).

The idea of using sports tourism for social, cultural, and community development has been promoted for decades. We previously mentioned that sports tourism achieves a significant impact on the creation of new value at the destination and that it can also have a significant impact on the employment of the population, on development and investments in infrastructure, as well as a very strong impact on the environment. However, certain authors (Weed & Bull, 2012) and especially Fredline (2005) rightly raise the question of strictly limiting and dividing the impact of sports tourism into economic, social, ecological, and others. As an example, he cites the impact on employment and the impact on infrastructure development. Is the impact on employment solely the economic impact and importance of sports tourism? If a larger number of people are able to perform activities and earn income from sports tourism, does that have any social implications? Do investments in infrastructure have only an economic dimension or primarily a social one? In what way will this infrastructure intended for the needs of sports tourism be accessible and used by the local population? How much and at what cost will the local population be able to use that infrastructure? In what way will it affect further urban development of the destination? What will be the impact on the entrepreneurial activity of the local population? What will be the impact on pollution and waste management? So obviously Fredline is right: It is hard to draw such strict parallels between multiple significances and impacts (Borovčanin & Lesjak, 2021). However, through scientific research papers and books, a kind of systematization can be created, while not denying the multidimensionality of each of the aforementioned influences.

The most often investigated social impacts in scientific works are as follows (Borovčanin & Lesjak, 2021):

• Impacts on the sociological structure of the population
• Impacts on the cultural and social values of the local community
• Outcomes of intercultural interaction
• Effects on the psychological well-being of the individual.

Especially the contribution to conflict resolution, reconciliation, and peace building is often analyzed and emphasized. In accordance with that and taking into account that the formulation of the already mentioned SDGs of the UN has been accepted and recognized as a catalyst for solving numerous social, economic, and environmental issues, below we will pay attention to the impacts of sports, tourism, and sports events that directly and indirectly contribute to the achievement of Goal 16 (promoting peaceful and inclusive societies for sustainable development, providing access to justice for all, and building effective, accountable, and inclusive institutions at all levels) and also to other goals (4, 8, and 11) that touch on sustainable social development.

Impact on the educational, cultural, and social values of the local community

Regular participation in sports and physical activities provides various social and health benefits. Not only does it have a direct impact on physical fitness, but it also encourages healthy lifestyle choices in children and young people, helping them stay active and fight noncommunicable diseases. Numerous studies conducted by the World Health Organization have also highlighted that physical exercise can stimulate positive mental health and cognitive development. Exercise has been linked to improved self-esteem and confidence, as well as positive effects on people struggling with depression and anxiety. In combination with the school curriculum, physical activities and sports are necessary for a comprehensive education. Sports provide lifelong learning and alternative education for children who cannot attend school. By participating in sports and physical activities alongside school, students are exposed to the key values of sports, including teamwork, fair play, respect for rules and others, cooperation, discipline, and tolerance. These skills are necessary for future participation in group activities and professional life and can stimulate social cohesion within communities and societies. Given the personal and social development benefits that sports offer, increasing access and participation is a primary development goal. For this reason, UNOSDP has been introducing the Young

Leadership Program (YLP) since 2012 with the aim of training and empowering young leaders from disadvantaged communities to use sports as a means of progress. At a YLP camp held in Hamburg, Germany, in February 2016, six refugees were welcomed and integrated into the group, highlighting the ability of sports to foster inclusion and bring people together (Lemke, 2016).

Education is also an integral part of the activities that accompany sports tourism, as well as the organization and implementation of sports events. The Olympic Games stand out as examples of good practice, which essentially represent the perfect platform from which to learn Olympic values – excellence, friendship, and respect. In Beijing, the organizing committee of the games, in cooperation with the Ministry of Education of China and the Chinese Olympic Committee, created a joint Olympic education program, which included 400 million Chinese children by integrating Olympic education into the existing curriculum in over 400,000 schools. The London 2012 Organizing Committee's Get Set education program saw 85% of schools across the country take part in a number of Olympic-related activities inspired by sports and the Olympic values (Committee, 2023).

In some cases, such as the Hong Kong Dragon Boat Festival, sports tourism events may have a cultural, religious, or even ritual association (Sofield & Sivan, 1994). According to Rinehart, the Super Bowl is "a modern pilgrimage ritual where people come to reunite with their friends and experience the event in concert with others of similar interest" (Rinehart, 1993). The aforementioned author's study of the 1992 Super Bowl in Minneapolis (Minnesota) showed that individuals attended the game to be seen, to support the team, and to continue the ritual. For many, the experience of a sporting event is a recurring, life-shaping experience (Rinehart, 1998).

In sports tourism research, the social impact was also analyzed through surveys of the local population about their perception regarding the specific event, manifestation, and the like. A study from 2011 that compared the perception of the local population before the organization of the famous cycling race "Tour de France" showed that the local public may have different perceptions regarding the impact and importance of this sporting event and the arrival of tourists for passive observation at their destination (Balduck et al., 2011). A good example of how a city, or in this case a country, can maximize exposure through a sporting event is Sydney, that is, Australia as the host of the 2000 Summer Olympics. Shortly after Sydney was awarded the right to host the Summer Olympics, the Australian Tourism Commission (ATC) prepared an Olympic Games tourism strategy and worked tirelessly to maximize every possible opportunity. Over 1,000 individual projects were implemented, and the results showed enviable success and great satisfaction of the local population (Hudson, 2003).

The significance of sports events and programs in the process of social inclusion

The holding of large sporting events is often presented as a tool for the development of urban communities (Misener & Mason, 2009; Clark & Misener, 2015), encouraging social inclusion (Collins et al., 2014), and reducing the crime rate

(Karaburn, & Balcioğlu, 2020). Lemke (2016) cites the Table Tennis project for NepALL as an example of the inclusion of people with disabilities, which showed how sports can foster social development by changing the perceptions about people with disabilities and giving such people the opportunity to play sports despite significant obstacles.

A large number of sports organizations and competitions deal with various social problems. Thus, the NBA has in its composition the "NBA Cares" program, which is an initiative of the league to deal with important social issues such as education, youth, family development, and health. The NBA and its teams have a number of programs, partners, and initiatives that strive to positively impact children and families around the world. "NBA Cares" includes more than 10,000 schools and community centers (NBA, 2023). The EuroLeague developed the "One Team" project, which gathers 16 clubs from nine countries; the eight teams that founded the project, Alba Berlin, Anadolu Efes Istanbul, CSKA Moscow, Maccabi Electra Tel Aviv, EA7 Emporio Armani Milano, Olympiacos Piraeus, Real Madrid, and Unicaja Malaga, joined the program in February 2012, while seven new clubs, Brose Baskets Bamberg, FC Barcelona Regal, Fenerbahce Ulker Istanbul, Galatasaray Medical Park, Panathinaikos Athens, Partizan mts Belgrade, and Union Olimpija Ljubljana, became members in November 2012. Each club works with a specific group of people who are excluded from their communities using the power of basketball as a means of integration (http://www.euroleague.net/one-team/members-projects, n.d.). Today, ten years later, a large number of clubs participate in this project and further develop and celebrate it in different ways: Milan (Italy) – involving students and their teachers, and Žalgiris (Lithuania) – with the help of the Association of Lithuanian Children's Day Care Centers, organize various lectures and sessions in the field – leadership, teamwork, responsibility, stress management, etc. (http://www.euroleague.net, 2023).

One of the latest studies, which included 24 studies that were eligible for a systematic review (13 of which were included in the meta-analytic integration), proved that participants in the observed sports programs showed a significant reduction in outcomes such as aggression or antisocial behavior. In the aforementioned research, psychological outcomes such as self-esteem or mental well-being were also analyzed, which also significantly improved when participating in sports programs (Jugl et al., 2021).

Sports tourism and events as a platform for mutual understanding, overcoming conflicts, and peace building

Although it is not uncommon to find limited empirical research that analyzes the strategic potential of sporting events in contributing to conflict resolution, reconciliation, and peace building in deeply divided societies, there are studies that define critical elements for promoting positive relations between communities, building local capacities, and improving overall social development. One of these deals with the experiences of numerous Football for Peace (F4P) projects that were implemented in Israel in 2009. In this study, strategies for managing sports events among different communities (Jewish, Arab, and Circassian) in the north of Israel were identified and investigated. Following an interpretive inquiry method, focus group discussions were conducted with key implementers and sports coaches who explored participants' experiences and used this information to develop practical recommendations for social development through organizing sports events. As a result, practical proposals have emerged that have transferable implications for other local organizations and nongovernmental organizations (NGOs) that use sporting events in divided and/or vulnerable communities elsewhere in the world (Schulenkorf & Sugden, 2011).

It is often pointed out that sporting events promote dialog, integration, and peaceful understanding among different groups, even when other forms of negotiation are not successful. However, in order for the social outcomes of sports events to not be only anecdotal, it is good to empirically examine the active engagement of groups with "others" in participatory sports events projects. A group of authors analyzed the potential of an intercommunity sporting event in contributing to intergroup development and building social capital in ethnically divided Sri Lanka. The results were interpreted based on the analysis of 35 in-depth interviews with Sinhalese, Tamil, Muslim, and international actors of the event. By providing evidence of different sociocultural experiences at the event, this research examines the impact of the event on intergroup relations and its impact on the stock of social capital available to communities. Findings can help governments, policymakers, and NGOs advance event-based policies and practices as vehicles and catalysts for improved intergroup relations and the creation of social capital (Schulenkorf et al., 2011).

A good example of how sports can be used to promote mutual understanding and dialog in conflict areas is the YLP held in Gwangju, Republic of Korea, in 2013. The program brought together participants from the Republic of Korea and the Democratic People's Republic of Korea, giving them and others the opportunity to realize that they share more similarities than differences, and helping them break down negative perceptions of each other. The YLP was an essential vehicle for the two countries to use sports to create social bonds that help foster rapprochement, respect, mutual understanding, and dialog (Lemke, 2016).

On the other hand, certain research deals with the processes that are necessary to understand the potential and possibilities of sports and sports events for building peace. By identifying initiatives in South Africa that are used at national, societal, and individual levels, Höglund and Sundberg present the possible effects of sports on reconciliation in divided states. By connecting experiences from state policies, activities of NGOs, and donor projects with social identity and the theory of reconciliation, the authors indicate the possible positive and negative aspects of sports. Finally, they suggest possible directions for further research to further identify the necessary measures to turn sporting events into effective political tools for postconflict peace building (Höglund & Sundberg, 2008).

It is obvious that due to their particularities, sports, tourism, and various types of sports events and happenings have a huge reach and international importance, so they rightly stand out as important for the achievement of the goals of the UN in the field of sustainable development and peace. As a case study and a good example of a sports event that, passing through various challenges, resisted various restrictions and continuously promoted messages of peace, below we present the Belgrade Marathon.

CASE STUDY: Belgrade Marathon – the race that reconciled the world

The Belgrade Marathon is the largest and most massive sports event in the Republic of Serbia. It started as an idea of a group of enthusiasts to renew a race that was run in Belgrade as far back as 1910, and the first marathon was officially held in 1988. Considering that marathon is one of the distinguishing features of the world's great metropolises, the organizers of this event (the sports association, the city of Belgrade, and all competent authorities) wanted to develop this event in the direction of the marathon becoming an international event that will gather in one place lovers of running from other countries who will be able to take part in different categories and ages.

Many famous names from the world of marathoners came to the Belgrade Marathon – Bob Beamon, Karl Lewis, Mike Puel, Simon Kipketer, and others. However, one name left a special mark and that is Fischel Lebowitz, better known internationally as Fred Lebow. He was an active runner, director and founder of the New York Marathon, director and founder of the New York Running Club, and a member of the US National Long-Distance Runners Hall of Fame. His influence in the world of running, especially street running and street racing, will be forever recorded in the history of sports and the world of running. His connection with Belgrade and the Belgrade Marathon began at a gathering in Los Angeles, where Mr. Dejan Nikolić, director of the Belgrade Marathon for many years, and Fred Lebow met by chance in 1988. Fred believed that running has no boundaries and belongs to everyone. He had a sincere desire to help the development of this race with his reputation and authority. Thus, in 1988, an acquaintance and friendship were created that lasted for years and whose unbreakable bond exists even today and after the death of Fred Lebow.

Due to the beginning of the conflict in the territory of the former SFRY, there were circumstances in which not only the organization of the marathon was called into question, but also the gathering of participants was restricted, and it became almost impossible to organize any event. Belgrade and SR Yugoslavia (at that time) were in difficult economic and political circumstances. During 1993, the country was gripped by galloping inflation, and on the international political level, sanctions were imposed on FR Yugoslavia. That same year (1993), it was almost unthinkable to organize an international sports event in Belgrade. And yet, the organizers of the Belgrade Marathon had an expressed desire to organize the event, sharing Lebow's attitude about the nonexistence of borders for sports and running. At the international conference of the World Association of Marathons and Street Races (AIMS) in Lisbon in 1993, Fred publicly took the position that Belgrade and the organizers of the Belgrade Marathon must be helped and that, despite everything, the race with international participation should be held. Despite the suggestions of

American officials that he should not go to Belgrade because the country is under sanctions, Fred first told everyone that he would go and participate in the Belgrade Marathon for that very reason, despite the significant health problems he was facing at the time. Fred Lebow, of course, came to Belgrade and participated in the Sixth Belgrade Marathon, and the information about Fred's presence at the Belgrade Marathon in 1993 went around the world. The prestigious world magazine "Runner's World" published a report about Fred Lebow's entire stay in Belgrade that year and made an article about his participation in the race.

A few months later, an author's text on Fred Lebow appeared on the front page of the *New York Times*, which conveyed the same message that Fred gave at the press conference in the Belgrade City Assembly. That message was also the reason for his stay in Belgrade – "Marathons do not tolerate any borders," and on that occasion, he called the Belgrade Marathon a marathon of "peace and hope." After the death of Fred Lebow, the Belgrade Marathon received permission to name its trophy the Fred Lebow Trophy as the only marathon in the world. This unique fact about this trophy is an expression of gratitude to Mr. Lebow for his immeasurable contribution to the largest international sports event that is continuously held in Belgrade.

Marathon is a sport stronger than the 1999 bombing in spreading the message of peace and hope

Not long after the death of Fred Lebow, the Belgrade Marathon faced a new challenge. Again, it was not a matter of sports challenges but challenges of international relations. And this time, the power of sports and international sports competitions proved to be stronger than international politics and international relations. The Belgrade Marathon was also held in 1999 during the bombing of FR Yugoslavia with two powerful slogans "Run for fun, not from bombs" and "Run for the world, stop the war" (or "Stop the war, run the world"). Due to the country's blockade and other circumstances, the race participants were welcomed at airports in neighboring countries and then brought to Belgrade. The unstoppable power of sports for the international community was demonstrated at that moment by Primo Nebiolo, the president of the World Association of Athletics Federations (IAAF, today World Athletics), who sent information to the world media that the Belgrade Marathon would be held on 17 April. This powerful message showed the impact of sports events that no one could resist and that no one could prohibit. Symbolically, the elite participants of the Belgrade Marathon crossed the finish line holding hands, so they were all winners, and this result was officially recognized internationally by the IAAF (today's World Athletics).

From the marking of the track, through the marathon of peace and hope and the Fred Lebow trophy, overcoming the bombing and new challenges, the Belgrade Marathon today represents one of the biggest sports events in Serbia, and it is particularly noteworthy that it is adapted to different categories and all ages:

- Children's marathon: A race intended for the youngest
- Pleasure Run: The event with the largest number of participants

- Half marathon: It is done according to all world standards, and it is intended for a more serious racing population
- Marathon: The main race of the event, intended for the most prepared runners.

All events on the streets of the city are followed by all domestic and a large part of foreign media. Reports from the event are broadcast by Eurosport, Sky Sport, and ISPN, and the presence of large media companies and the publicity that the event receives provide the opportunity for the whole world to participate in the marathon in this way and greet the winners and friends of the event.

Conclusion

Most of the research on the social impact of sports tourism and events, regardless of methodological differences and limitations, was done on examples, that is, case studies from which it is difficult to generalize and draw one conclusion. The scope of the social impact is very heterogeneous, and the benefits for society and the overall development of the local community are multiple. In this chapter, we have shown that sports, tourism, and events, as a whole, represent a powerful platform for mutual understanding, fostering peace and developing societies with respect and appreciation for the diversity that exists around the world. Sports, tourism, and events have a unique power to accelerate the process of rebuilding post-conflict societies and uplifting individuals and communities. By developing and promoting sports tourism and events, a good basis is created for fostering respect, solidarity, and reconciliation with interaction between different communities that lead to better mutual understanding.

Bibliography

Aho, S. K. (2001). Towards a general theory of touristic experiences: Modelling experience process in tourism. *Tourism Review*, 33–37.

Bartoluci, M. (2004). *Menedžment u sportu i turizmu*. Zagreb: Ekonomski fakultet.

Bianchi, R. V., & Stephenson, M. L. (2013). Deciphering tourism and citizenship in a globalized world. *Tourism Management*, 10–20.

Borovčanin, D., & Lesjak, M. (2021). *Sportski turizam*. Beograd: Univerzitet Singidunum.

Borovcanin, D., & Lesjak, M. (2022). Sport tourism participants. In *Encyclopedia of tourism management and marketing*. Cheltenham Glos, UK: Edward Elgar Publishing.

Butler, R. (2015). The evolution of tourism and tourism research. *Tourism Recreation Research*, 16–27.

Čerović, S., Brdar, I., & Knežević, M. (2022). *Turizam Teorija I Principi*. Beograd: Univerzitet Singidunum.

Čerović, S., Knežević, M., & Pavlović, D. (2016). The effects of tourism on the GDP of Macedonia, Montenegro and Serbia in the process of European integration. *Amfiteatru Economic Journal*, 407–422.

Čerović, S., Knežević, M., Matović, V., & Brdar, I. (2015). The contribution of tourism industry on the GDP of Western Balkan countries. *Industrija*.

Clark, R., & Misener, L. (2015). Understanding urban development through a sport events portfolio: A case study of London. *Journal of Sport Management*, 11–26.

Cohen, E. (1979). A phenomenology of tourist experiences. *Sociology*, 179–201.

Collins, M. F., Collins, M., & Kay, T. (2014). *Sport and social exclusion*. London: Routledge.

Collison, H., Darnell, S., Giulianotti, R., & Howe, P. D. (2017). The inclusion conundrum: A critical account of youth and gender issues within and beyond sport for development and peace interventions. *Social Inclusion*, 223–231.

Committee, I. O. (2023, March 15). *www.olympic.org/Documents/Olympism_in_action/ Legacy/2013_Booklet_Legacy.pdf*. Retrieved from https://olympics.com/; www.olympic. org/Documents/Olympism_in_action/Legacy/2013_Booklet_Legacy.pdf

Erfurt, R. A., & Johnsen, J. (2003). Influence of an event on a destination's image – the case of the annual meeting of the World Economic Forum (WEF) in Davos/Switzerland. *Tourism Review*, 21–27.

www.euroleague.net/one-team/members-projects. (n.d.). Retrieved from www.euroleague. net;www.euroleague.net/one-team/members-projects

www.euroleague.net. (2023). Retrieved from www.euroleague.net;www.euroleague.net/ one-team/news/i/123754

Fernández Gavira, J., Huete García, M. Á., & Vélez Colón, L. (2017). Vulnerable groups at risk for sport and social exclusion. *Journal of Physical Education and Sport*, 312–326.

Fredline, E. (2005). Host and guest relations and sport tourism. *Sport in Society*, 263–279.

Gammon, S., & Robinson, T. (2003). Sport and tourism: A conceptual framework. *Journal of Sport & Tourism*, 21–26.

Getz, D. (2008). Event tourism: Definition, evolution, and research. *Tourism Management*, 403–428.

González-García, R. J., Mártínez-Rico, G., Bañuls-Lapuerta, F., & Calabuig, F. (2022). Residents' perception of the impact of sports tourism on sustainable social development. *Sustainability*, 1232.

Gozalova, M., Shchikanov, A., Vernigor, A., & Bagdasarian, V. (2014). Sports tourism. *Polish Journal of Sport and Tourism*, 92–96.

Higham, J. (1999). Commentary-sport as an avenue of tourism development: An analysis of the positive and negative impacts of sport tourism. *Current Issues in Tourism*, 82–90.

Higgins-Desbiolles, F. (2006). More than an "industry": The forgotten power of tourism as a social force. *Tourism Management*, 1192–1208.

Hinch, T. D., & Higham, J. E. (2001). Sport tourism: A framework for research. *International Journal of Tourism Research*, 45–58.

Höglund, K., & Sundberg, R. (2008). Reconciliation through sports? The case of South Africa. *Third World Quarterly*, 805–818.

Hollinshead, K., & Xiao Hou, C. (2012). The Seductions of 'Soft Power': The call for multi-fronted research into the articulative reach of tourism in China. *Journal of China Tourism Research*, 227–247.

Hudson, S. (2003). *Sport and adventure tourism*. London: Routledge.

Hussain, A. (2021). A future of tourism industry: Conscious travel, destination recovery and regenerative tourism. *Journal of Sustainability and Resilience*, 1(1).

Jiménez-García, M., Ruiz-Chico, J., Peña-Sánchez, A. R., & López-Sánchez, J. A. (2020). A bibliometric analysis of sports tourism and sustainability (2002–2019). *Sustainability*, 2840.

Judd, D. R., & Fainstein, S. S. (1999). *The tourist city*. New Haven, Connecticut: Yale University Press.

Jugl, I., Bender, D., & Lösel, F. (2021). Do sports programs prevent crime and reduce reoffending? A systematic review and meta-analysis on the effectiveness of sports programs. *Journal of Quantitative Criminology*, 1–52.

Karaburn, M., & Balcioğlu, İ. (2020). Sports as a crime prevention instrument: A mini review. *Yeni Symposium*, 58(4).

Larsen, J. (2019). Ordinary tourism and extraordinary everyday life: Rethinking tourism and cities. In C. S. Thomas Frisch (Ed.), *Tourism and everyday life in the contemporary city* (pp. 24–41). London: Routledge.

Lemke, W. (2016). The role of sport in achieving the sustainable development goals. *UN Chronicle*, 53(2), 6–9.

MacCannell, D. (2013). *The tourist: A new theory of the leisure class*. Oakland, CA 94607: University of California Press.

Mato, B., & Nevenka, Č. (1998). *Turizam i sport*. Zagreb: Fakultet za fizičku kulturu.

Misener, L., & Mason, D. S. (2009). Fostering community development through sporting events strategies: An examination of urban regime perceptions. *Journal of Sport Management*, 770–794.

Mollah, M. R., Cuskelly, G., & Hill, B. (2021). Sport tourism collaboration: A systematic quantitative literature review. *Journal of Sport & Tourism*, 3–25.

Morfoulaki, M., Myrovali, G., Kotoula, K. M., Karagiorgos, T., & Alexandris, K. (2023). Sport tourism as driving force for destinations' sustainability. *Sustainability*, 2445.

Morgan, H., Bush, A., & McGee, D. (2021). The contribution of sport to the sustainable development goals: Insights from Commonwealth Games Associations. *Journal of Sport for Development*, 14–29.

Mossberg, L. L. (2000). *Evaluation of events: Scandinavian experiences*. New York.

NBA. (2023, March 10). *https://cares.nba.com/socialimpactreport/*. Retrieved from https://cares.nba.com/; https://ak-static.cms.nba.com/wp-content/uploads/sites/54/2022/08/NBA22_ESG_00_Full_FINAL_digital.pdf

Pedauga, L. E., Pardo-Fanjul, A., Redondo, J. C., & Izquierdo, J. M. (2022). Assessing the economic contribution of sports tourism events: A regional social accounting matrix analysis approach. *Tourism Economics*, 599–620.

Pedersen, S. B. (2020). A passport to peace? Modern tourism and internationalist idealism. *European Review*, 389–402.

Pivac, T., & Stamenković, I. (2011). *Menadžment događaja*. Novi Sad: Prirodno-matematički fakultet u Novom Sadu.

Rinehart, R. E. (1993). *"Been there, did that": Contemporary sport performances*. Champaign, Illinois: University of Illinois at Urbana-Champaign.

Rinehart, R. E. (1998). *Players all: Performances in contemporary sport*. Bloomington, Indiana: Indiana University Press.

Schulenkorf, N., & Sugden, J. (2011). Sport for development and peace in divided societies: Cooperating for inter-community empowerment in Israel. *European Journal for Sport and Society*, 235–256.

Schulenkorf, N., Thomson, A., & Schlenker, K. (2011). Intercommunity sport events: Vehicles and catalysts for social capital in divided societies. *Event Management*, 105–119.

Sofield, T. H., & Sivan, A. (1994). From cultural festival to international sport – the Hong Kong dragon boat races. *Journal of Sport Tourism*, 5–17.

Stylidis, D., Woosnam, K. M., & Tasci, A. D. (2022). The effect of resident-tourist interaction quality on destination image and loyalty. *Journal of Sustainable Tourism*, 1219–1239.

Telfer, D. J., & Sharpley, R. (2015). *Tourism and development in the developing world*. London: Routledge.

Tsekouropoulos, G., Gkouna, O., Theocharis, D., & Gounas, A. (2022). Innovative sustainable tourism development and entrepreneurship through sports events. *Sustainability*, 4379.

United Nations General Assembly. (2015). *Transforming our world: The 2030 Agenda for sustainable development*. Retrieved from https://sustainabledevelopment.un.org/post2015/transformingourworld

Unković, S., & Zečević, B. (2016). *Ekonomika turizma*. Beograd: Ekonomski fakultet.

Van Rheenen, D., Cernaianu, S., & Sobry, C. (2017). Defining sport tourism: A content analysis of an evolving epistemology. *Journal of Sport & Tourism*, 75–93.

Wang, N. (1999). Rethinking authenticity in tourism experience. *Annals of Tourism Research*, 349–370.

Weed, M., & Bull, C. (2012). *Sports tourism: Participants, policy and providers*. Warsaw, Poland: Routledge.

Zagnoli, P., & Radicchi, E. (2009). Do major sports events enhance tourism destinations? *Physical Culture and Sport. Studies and Research*, 44–63.

Zhuang, X., Yao, Y., & Li, J. (2019). Sociocultural impacts of tourism on residents of world cultural heritage sites in China. *Sustainability*, 840.

Index

For Product Safety Concerns and Information please contact our EU
representative GPSR@taylorandfrancis.com
Taylor & Francis Verlag GmbH, Kaufingerstraße 24, 80331 München, Germany

www.ingramcontent.com/pod-product-compliance
Lightning Source LLC
Chambersburg PA
CBHW060311220326
41598CB00027B/4303

9 781032 471501